D. Leonard
Corgan
Library
COLLEGIUM · CHRISTI · REGIS
A·D·1946
KING'S
COLLEGE
WILKES-BARRE, PENNSYLVANIA

Tall Ships

A superbly illustrated and definitive guide to the world of sail training with particular attention to the ships themselves.

Sail training was a common feature of the nineteenth and early twentieth centuries, but as steam replaced sail for cargo-carrying and defence, navies and ship owners reduced or paid off their sail training vessels. This left a significant but reduced world fleet which today is beginning to expand again, as governments, companies and educationalists see the contribution that training under 'canvas' can make to an understanding of the environmental elements of the oceans.

TALL SHIPS is divided into three sections: the past, the present and the future—for tall ships are still being built for sail training. Each section contains an illustrated introduction followed by descriptions and photographs of each of the thirty-nine ships featured, twenty-one of which have been highlighted by Mike Willoughby's superb full-colour technical illustrations. There is an appendix of ships' rigging, full technical data on each vessel, and a bibliography.

TALL SHIPS is the official publication of the Sail Training Association.

ship in 1927 and had extensive experience under sail before the Second World War. From 1942 to 1945 he was a submarine captain, and since then has been Master of various sailing vessels, including *Sir Winston Churchill* and *Regina Maris*. He was the designer and rigger of *Captain Scott*, and has designed a 3,000 ton three-masted barque for the Greek Navy. His wide knowledge and experience of tall ships has made him much sought after as a technical adviser for such organisations as the Maritime Trust and the B.B.C.

TALL SHIPS

The World of Sail Training

Text by Maldwin Drummond
Illustrated by Mike Willoughby
Foreword by H.R.H. the Duke of Edinburgh

The Official Publication of the Sail Training Association

G. P. Putnam's Sons

G. P. Putnam's Sons
First American Edition 1976

Library of Congress Catalog Card Number: 76–394
SBN: 399–11751–2

Set in Monotype No. 7 by
Morrison & Gibb Ltd, London and Edinburgh
Printed in Italy

PHOTO CREDITS

Associated Press: p. 37 (2); Balcomb, K.: p. 113 (2); Beken & Sons: pp. 47 (1, 2), 48, 133 (1), 144; die Bildstelle der Marine: p. 131 (2); Birchall, Clifford: p. 33 (2); Bromley-Martin, Angela: endpapers, pp. 70, 94 (1, 2, 3, 4), 95 (1, 2), 97 (1), 126, 143 (1); Burns, Michael: p. 10; Central Press Photos: p. 121 (2); Christoffersen, A.: p. 30; Derby, W. L. H.: pp. 28 (1, 2, 3), 29 (1); Drummond, Maldwin: p. 148 (2); Dyevre, Capitaine de Vaisseau: pp. 114, 117; Fadda, Capt. Renato: pp. 58, 59 (1, 2, 3); Falgreen Ltd: p. 52; Goddard, Robert: p. 35 (2); Greenway, the Hon. Ambrose: p. 65 (2); Howard, Lt. Cmdr. the Hon. Greville, R.N.: p. 98; Ikeda, Capt. Isao: pp. 122, 125 (1, 2, 3); Imperial War Museum: pp. 11 (1, 2), 46 (2); Laivateollisuus Ab, Oy: p. 44 (3); Lewandowski, Zbigniew: pp. 46 (1), 50 (1), 135 (1); Loch Eil Trust: p. 85 (1, 2, 3); Lunch, M.: p. 93 (1); Lund, Kaj: pp. 45 (2), 55 (1), 65 (1), 69 (2), 121 (1); May, Capt. J. H. Le, Chilean Navy: p. 132; Meulen, Cees van de: p. 105 (2); Mircea Naval School, Constanta, Romania: pp. 138, 139 (1, 2, 3, 4); Moray Sea School: p. 13 (2); National Maritime Museum: pp. 9 (1, 2), 14, 15 (1, 2, 3), 19 (1, 2), 20, 23 (1, 2, 3, 4), 27 (1), 29 (3), 33 (3), 45 (1), 140; Nationale Vereniging 'Het Zeelend Zeeschip': pp. 102, 105 (1, 3); Navy League: p. 89 (2, 3, 4); Osborn, G. A.: pp. 66, 97 (2); Pamir/Passat Association: pp. 33 (1), 42; Photomasters: p. 44 (2); Prölss, Professor W.: p. 141 (2); Royal Danish Embassy, London: p. 55 (2, 3); Royal Swedish Navy: pp. 74, 77 (1, 2, 3); The *Scotsman*: p. 13 (1); Sea Education Association, Boston: p. 113 (1); Skyphotos: p. 64; Smith, Roger: p. 89 (1); Statens Sjohistoriska Museum, Sweden: pp. 12, 34, 35 (1); Staubo, Jan: p. 73 (1); Thorsen, Capt. Kjell, Royal Norwegian Navy: p. 69 (1, 3); *Times* Newspapers, London: p. 86; University of Southampton: p. 37 (1); U.S. Coast Guard Academy: pp. 31 (1, 2), 109 (2), 110 (1, 2, 3), 111 (1, 2); Villiers, Alan: pp. 19 (1, 2), 109; Warnholtz, Rolf: p. 101 (1, 2); *Western Morning News*, Plymouth: p. 29 (2); Willoughby, Mike: p. 81 (1, 2); Wundshammer, Benno: pp. 60, 63 (1), 77 (4), 111 (3), 129 (1, 2), 131 (1, 2, 3), 134; Yacht Klub Morski, Gdynia: p. 50 (3); *Yachts and Yachting*: pp. 78, 81 (3, 4), 136, 137 (1, 2); *Yachting World*: p. 101 (3).

Contents

Foreword

by H.R.H. the Duke of Edinburgh

Sail training has been a source of controversy among seamen ever since power driven ships came into use. Training in sail is not an essential part of the technical training of officers and crewmen of modern ships. However, experience in sail is a highly desirable part of the training of men. The wind and the sea are unforgiving elements and it takes skill and courage to live with them and to use them to advantage. The self-discovery and self-confidence which comes from training under sail are life-long assets.

This book reviews the whole story of training at sea and the ships in use. It is not based on nostalgia for a departed age or regret for a romantic way of life. The book is concerned with the future and with the young. It makes it quite clear that in this day of mechanical convenience and of the restrictions of urban industrial societies, training under sail can give the human spirit a real chance to develop and to mature.

Author's introduction

This book is focused on ships: their purposes, their ways and their achievements. In order to write about them, it has been necessary to draw on the experience, enthusiasm and devotion of those who have been interested in, or concerned with, these fine vessels, and I would like to thank here all those who have helped me. They have not only provided a wealth of information and photographs, but a fair measure of encouragement.

The Sail Training Association has played no small part in this. I am particularly grateful to our Patron, H.R.H. Prince Philip, who has written the Foreword. I know from my own time as Chairman of the Association, the great measure of encouragement and help that His Royal Highness has given to the organisation and I think it may fairly be said that, without his advice and energy, the Association would not have flourished as it has done.

Hugh Goodson, the President of S.T.A., introduced me to the world of training ships, after I had written an appreciation of how the three masted iron schooner *Result*, now owned by the Ulster Folk Museum but then for sale, could be used for a modified and minor form of classic sail training. My idea was to design an adventure training course around her traditional employment – that of trading between the north coast of France and the southern half of Britain and Ireland. The cargo I selected was one that was not too difficult to place – barrels of *vin ordinaire*.

However, before I had a chance to bid for the ship, Hugh Goodson persuaded me to turn my attentions to the idea of building a new adventure training ship for Britain, and no doubt he saved me from a lifetime of arguments with Customs officials and inspectors from the British Board of Trade. Subsequently, the Council of the Association made me Chairman of the new technical committee set up to mastermind the building of the three masted topsail schooner *Sir Winston Churchill*. Their confidence carried me through to the Chairmanship of the Association and provided a vantage point from which to admire the work of all those who harnessed the natural advantages of both sea and wind, whether for training for the sea as a profession, or for life itself, through adventure schemes. As a result I was able to work with and meet those who had spent a lifetime under sail and those who were engaged in training others for the sea. I refer particularly to Commander Alan Villiers, Alex Hurst and Captain Whally Wakeford, who have given me information, photographs and advice of great value.

Returning to the Sail Training Association, I must single out and thank Col. Dick Scholfield, until 1976 the Sail Training Association's race director. He has read the manuscript and his help has been invaluable as has that of Lord Burnham, Vice-Chairman Sailing (U.K.). His successor elect, John Hamilton, has supplied his own individual view of sail training and this is based on experience with the *Rona* scheme and as mate on the *Captain Scott*. Captains Tony Stewart and Chris Phelan, who followed Captain Whally Wakeford as principals of the Southampton School of Navigation, have also contributed their views and comments for which I am grateful. Captain David Bromley-Martin, the Sail Training Association's schooner director and that committee's chairman, Brian Stewart, have also given valuable help.

Admiral Sir Charles Madden, a past chairman of the Sail Training Association sailing committee, and now of Britain's National Maritime Museum, Greenwich, and Basil Greenhill, the museum's director, have also given aid and encouragement. I am particularly grateful to the museum for the generous use of their photographic library, to the Photographic Librarian, G. A. Osbon, and to Mrs. Margaret Johnson, who helped me with research.

I must also mention Freddie Cartwright, who championed the cause of the schooners *Sir Winston Churchill* and *Malcolm Miller* in Wales during my chairmanship and who succeeded me – he has been an inspiration through his ideas on adventure training vessels of the future; Colin Mudie, who was responsible for the complicated formula that enables the S.T.A. to race 'out and out' ocean racers against large square-rigged vessels and who is one of Britain's most adventurous naval architects, gave me the benefit of a look into his crystal ball by examining ships to come; Professor Prölss from the Department of Naval Architecture at Hamburg University, who provided me with a similarly stimulating experience, showing how lessons of the past can be wedded to modern technology to provide a future for trading under sail; Lt. Cdr. the Hon. Greville Howard, the S.T.A.'s vice-chairman sailing (Overseas), who also helped, being an expert photographer; and, indeed, I would like to thank all members of the Council of S.T.A. for their support and interest.

Captain Hans Engel, captain of the *Gorch Fock*, and a vice-patron of the Sail Training Association, has given me invaluable help too. Few people know more about the advantages of training naval officers under sail than he. His contribution to sail training has been very considerable and was crowned by his part in organising Operation Sail, Kiel, in 1972. Captain Espanak, too, a past master of the *Christian Radich*, supplied me with the benefit of his experience when the idea of a book was first proposed and this helped form my ideas for the layout. Barclay Warburton III, a founder of the American Sail Training Association, has also given me important information.

My researches were made both easier and considerably more rewarding by the Liverpool-based magazine *Sea Breezes*, for in response to my letter in April of 1975, I received page upon page of helpful information from all around the world. I thanked all those who wrote personally, especially Fred G. Shaw, but would like to record again here my thanks to them and my gratitude to the editor of *Sea Breezes*, Craig J. M. Carter who, by publishing the letter, launched this remarkable response.

I have included in the list below, not only those who replied to my letter in *Sea Breezes* but, I hope, all those who helped me with advice.

Aboe, Lt. Col. (Naval Attaché, Indonesian Embassy, London); Abramson, Erik (Mariners International); Adams, Lionel; Alan-Williams, Cdr., K.R.N.; Amarandidis, Vice-Adm., C.H.N. (P.O.C., Commander in Chief, Ministry of Mercantile Marine, Piraeus); Anderson, A. O.;

Anthony, M.; Ayre, Colin C.; Balfour, The Earl of; Bedford, C. H.; Bendt, Kurt; Berge, Dr. Yur Louis Jh; Bergmann, Finn (Directoratet for Sofartsuddanelsen, Copenhagen); Birchall, Clifford; Bjerke, Alf (S.T.A.'s representative in Norway); Bonds, Lt. Cdr. John B., U.S.N. (Rear Commodore U.S.N.S.A.); Brock-Davis, Annette; Cammell, Mrs. (Schooner Secretary, S.T.A.); Christoffersen, Capt. A.; Cobb, Cdr. David, R.N. (Chairman, Asscn. of Sail Training Organisations); Collis, Capt. Patrick; Cook, Peter; Corlett, Euan (Burness, Corlett & Partners); Corujo, Capt. P. M. Guerra, Po.N. (Naval Attaché, Portuguese Embassy); Cowp, R. W.; Cramer, Corwith (Executive Director, Sea Education Asscn., Boston); Cross, D. J. (Senior Architect, The Archives, Authority of New South Wales); Crumlin-Pedersen, Ole (Director of the Institute of Maritime Archaeology, National Museum of Denmark and Custodian of the Viking Ship Museum, Roskilde); Currie, John; Davenport, Montague; Derby, W. L. H.; Dixon, Robert (Asst. Librarian, U.S. Coast Guard Academy, New London, Conn.); Drummond, Aldred; Dusa, Capt. 1st Rank A. Ro. N. (Naval Attaché, Romanian Embassy, London); Dyevre, Capitaine de Vaisseau (Le Commandant, Ecole Navale et Militaire de la Flotte); Engelsen, Finn; Eugen, Capt. 3rd Rank Ispas (Mircea Naval School, Constanta, Romania); Eyre-Walker, Miss M.; Fadda, Capt. Renato (Stato Maggiore della Marina, Rome); Fairchild, Mrs. Flora (co-ordinator of Development Projects, Mystic Seaport Inc.); Fairley, Gordon (Royal Yachting Association); Feather, John M. (Warden, Outward Bound Moray Sea School); Fischer, Herr I. V. (Gesellschaft fur Sport und Technik, Berlin, German Democratic Republic); Gay, Cdr. David, R.N.; Goddard, Robert H. I.; Gould, Andrew (R.N.L.I.); Glass, W.; Grant, Capt. W. C.; Greenway, The Hon. Ambrose; Griffiths, Capt. Glyn; Groenhart, Nicholas; Hamilton-Meikle, Capt. R. N.; Hammerstrom, Dr. Carl Frederick; Hansen, Miss Birthe; Hansen, Capt.; Hansen, Hans Jurgen; Hawkins, Richard (Associate Director, Sea Education Asscn., Boston); Hayman, Bernard (*Yachting World*); Healy, Capt. C. F. (Director, Coiste an *Asgard*, Dublin); Hedlund, Commodore Bengt, R.Sw.N.; Hein, Cdr. Jasper, R.Sw.N.; Herzogin Cecilie Museum, Salcombe; Humphreys, Arthur (co-ordinator, The Oceanic Society); Ikeda, Capt. Isao (Director-General Training Dept., Institute for Sea Training, Ministry of Transport, Tokyo); Irwin, Capt. James C. (U.S. Coast Guard Academy, Commanding Officer, *U.S.C.G.C. Eagle*); Jespersen, Capt. B. Barnen; Johnson, A. R. (Director of External Affairs, Outward Bound Trust); Johnson, Paul H. (Librarian Curator, U.S.C.G. Academy, New London); Juta, Cdr. H. M., R.N.N.; Kennedy, Alstan; Kok, C. S.; Kolesnik, Eugene M. (editor *Janes Yearbooks*); Kowalezyk, Rektor, Wyzsza Szkola Morska, Gdynia; Laabs, Cdr. P. H. H. (C.N. Asst. Naval Attaché, Federal Republic of Germany, London); Laivateollisuus, Ab, Oy; Le May, Capt. J. H., Ch.N. (Deputy Chief of the Chilean Naval Mission, Chilean Embassy, London); Lund, Kaj (Danish Merchant Navy Welfare Board, Copenhagen); Mackenzie, John (Herd & Mackenzie); Maddocks, Miss P. M. (Photographic Librarian, U.S.N. Institute); Margot, Enrique; Matz, Erling (Statens Sjohistoriska Museum); Mitchell, F. C. (Falkland Isles Co. Ltd., London); Mills, Tony (Cinemarine, Bexhill); Moberg, Eihar; Munro, Barry R.; Needham, Jack; Nelson, Maj. Gen. Sir John, K.C.V.O., C.B., D.S.O., M.C. (President, Asscn. of Sail Training Organisations); The News Centre, *Portsmouth Evening News*; Nicholson, Christopher (Camper & Nicholson, Southampton); Nolan, Capt. C. J. (Director of Administration and Commanding Officer of T.S. Empire State IV Maritime College, State University of New York); Norrsell, Cdr. Lars, R.Sw.N. (Chief of Naval Training Dept., Royal Swedish Navy); Paizis-Paradelis, Capt. C. (Naval Attaché, Hellenic Embassy, London); Pamir/Passat Association (Karl Gerisch); Headmaster of Pangbourne; Parker, Cdr. Michael; Patch Willoughby, Mrs.; Pearce, A. C. B.; Picard, Henri; Pogdanovic, Capt. N. Radomir (Yugoslav Naval Attaché, Yugoslav Embassy, London); Ponselle, Capitaine de Fregate (Asst. Naval Attaché, French Embassy, London); Powell, Joe (Cinemarine, Bexhill); Reksten, Hilmar; Reynolds, J.; Schmidt, John Host (Press and Cultural Attaché, Royal Danish Embassy, London); Simpson, A. W.; Staubo, Jan; Stewart, Capt. Tony; Stanton, Gerald; Sturges, Michael D. (Director of Education, Mystic Seaport Inc.); Summers, Sir Spencer (Chairman, Outward Bound Trust); Swedish Naval Attaché (Swedish Embassy, London); Swindells, Capt. Barry; Symes, E.; Taylor, Dr. B.; Tetzlaff, Mrs.; Thomson, James B.; Thomson, S.; Thorsen, Capt. Kjell, R.No.N.; Turff, N. S.; Tyrrell, John (Tyrrells of Arklow); Van Dam, Kaes; Wall, Raymond (Camper & Nicholson, Southampton); Vanderstar, Cornelius; Ward, H. (Borough Librarian, London Borough of Tower Hamlets); Williams, Geoffrey (Director of the Ocean Youth Club); Zaratiegui, Capitain de Fragata Horacio, At. N. (Dept. of Information, Argentine Navy).

I have reserved particularly the last paragraph and the last words of thanks for those who have worked alongside me, without whom this book would not have been possible.

I owe a considerable debt of gratitude to Captain Mike Willoughby, who did the drawings. He tells me he was aided in his researches by plans lent or given by owners, by designers, and by builders. I would like to echo this and add my further thanks to the owners of vessels who filled in the forms I sent them. He checked his information where appropriate with the drawings contained in the invaluable work of Harold Underhill, *Sail Training and Cadet Ships*, which I have referred to later and in the bibliography.

The work of collecting the photographs and masterminding their selection, was done by Angela Bromley-Martin, the photographic librarian of Sail Training Association, who has considerable ability in this direction. Certainly, no one could have undertaken the task with greater enthusiasm and I would like to pay tribute to her. My thanks go also to Richard Vine, who was responsible for the layout.

Finally, I would like to thank Mrs. Rosalie Hendey, who typed almost everything several times over, and must by now have collected as particular a view of the sea as that watched over by her pilot service husband.

Ships of the past

'A passage under sail brings out in the course of days whatever there may be of sea love and sea sense in an individual whose soul is not indissolubly wedded to the pedestrian shore.'

Joseph Conrad aboard *Torrens*

Conrad put into a few words the idea of the training gained from a sailing ship. Learning by doing has been part of the tradition of man's mastery of the ocean environment, since water has been used for transport, defence and aggression. Would-be sailors learned through experience, aboard trading and fighting vessels of ancient Egypt and Greece, aboard the fleets of Alfred the Great beset by the Danes, and in the long ships of the Vikings themselves. This process continued with little challenge until the 19th century. It included the scheme developed on naval lines by the East India Company, which trained its own midshipmen aboard vessels already carrying cargo.

The need for a vessel built or adapted specifically for training is a comparatively new idea. The concept produced two main types of vessel. The first of these was the sail training vessel using the term in its classical form, the second was the cadet ship.

Sail Training Vessel – Classical Form

The classical form of sail training vessel was a sailing ship carrying a comparatively large number of boys (acting as midshipmen), such as premium paying apprentices who were potential officers, and boys who would eventually become ordinary seamen and A.B.s. By tradition, the ship carried large amounts of cargo and a number of passengers. The trainees were additional to the normal ship's complement and schoolmasters, as well as other staff, were carried for their instruction. This idea was sometimes stretched to include ships that carried apprentices who were really nothing more than cheap labour. Training while trading summarises the role of the classical sail training ship.

The Cadet Ship

These were sailing ships that carried no cargo, only trainees, and were for the most part run for the benefit of the Navy. Among the exceptions were two American training vessels, the ship-rigged corvette *Saratoga* and the ship-rigged, 20 gun sloop *St. Mary's*, run by the cities of Philadelphia and New York respectively. The cadet ship gave more than sea experience – it provided a naval education afloat and carried only trainees. In this book the term 'cadet ship' embraces the school ship.

Adventure Training Ships

There came to be a third type, the adventure training ship. The beginnings of this idea are difficult to plot, for there is no doubt that the classical sail training vessel and perhaps the cadet ship, carried those who were not destined to make the sea their career but saw in it a training for life. It can, therefore, be said that both the sail training vessel and the cadet ship played a parental role in the birth of the adventure training vessel. Certainly Alan Villiers, with his *Joseph Conrad* (see page 16), can claim to be one of the first to experiment with the idea. The concept was given further strength by the Outward Bound's *Prince Louis I* (ex *Maisie Graham*).

In order to illustrate the development and the difference between these three types of ships and their training methods, it is worth looking at the vessels themselves. The divergence, of course, continues today, as shown later, though now the three types of vessel are grouped under the common name 'sail training vessel' borrowed from the classical form. The photographs below illustrate two of the three types of ship.

1. Port Jackson: the classical sail training ship —training while trading.

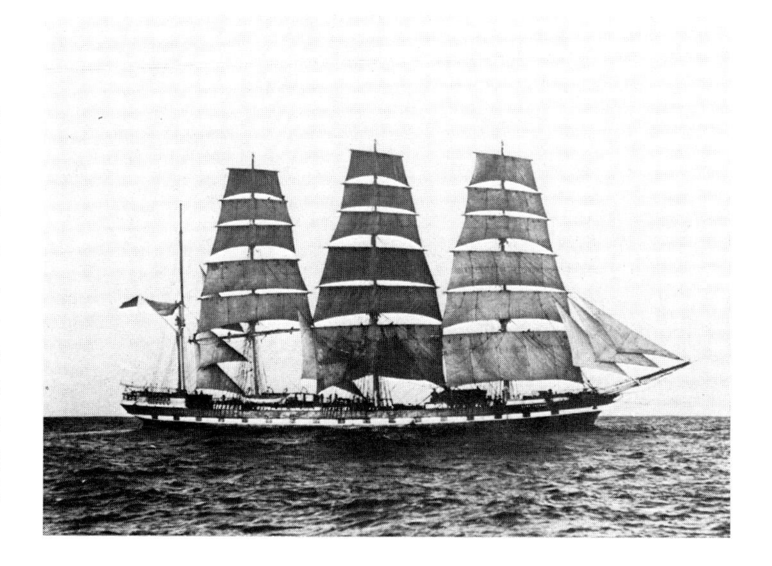

2. Joseph Conrad: an adventure training ship—training through the sea not necessarily for the sea.

Classical Sail Training

The term 'sail training' belongs rightly to the cargo/cadet ships of the last half of the 19th and the first half of this century. These vessels were primarily merchantmen and were built for a specific trade, carrying mainly grain from Australia or nitrate from the west coast of South America. The cadet or apprentice principle was built into them. The discipline of the ships' trade took them, as a matter of course, into the high latitudes, round the Horn and through sea areas traditionally characterised by strong winds, seas created by endless fetch, and the ability of damp to penetrate everywhere. These conditions are avoided by modern cadet ships whose programmes are usually designed to accommodate less demanding circumstances, and it was this trading discipline, coupled with the time that it took to achieve a round passage, that gave such reputation to those who had been trained in such conditions. As Alex Hurst remarks in an article in *Sea Breezes*: 'As a training to become an officer in a sailing ship, the sailing ships were excellent, but they rather lost their point when there were no sailing ships requiring officers. There were other advantages. A boy was taught a certain self-reliance when, over and over again he would be working (perhaps aloft) on his own.'

'The very nature of the life, particularly in high latitudes, not only taught him this, but also the way to get on with his fellows on long voyages and in trying conditions. If the knowledge of sail handling became of lesser consequence as the years rolled on, men who had served on such ships had gained a better appreciation of the very run of the sea and, indeed, of the very relationship between man and the elements.'

The latter part of Alex Hurst's last sentence applies, in varying degrees, to the whole ethic of working a ship under sail, whether a cadet, sail training, or adventure training vessel. The sail training ship, because of the conditions and experiences referred to, did it better.

An early example of the classical sail training ship was the *Macquarie* (ex *Melbourne*, see page 13), built in 1873 by R. H. Green of Blackwall, London, who were responsible for the famous and distinctive line of Blackwall frigates. In 1888, this full-rigged ship of 1,857 tons displacement was bought by Devitt & Moore of London for their training scheme. The idea was to combine passenger and cargo carrying to Australia with premium paying cadets, who would be educated by the ship's schoolmaster and at the same time learn the arts of the mariner at first hand. *Macquarie* and the similarly rigged *Illawarra* had succeeded the ship-rigged vessels *Harbinger* and *Hesperus* which had pioneered the scheme. In turn they gave way to the four masted barques *Medway* and *Port Jackson* (see page 38), until the commercial failure of trading under sail made such a scheme unworkable in 1921. By then, due to the twin disasters to *Medway*, converted to a tanker by the Navy, and *Port Jackson*, torpedoed in 1917, Devitt & Moore had already changed their pattern, using the school ship *St. George* for pre-sea training.

The last line of British sailing ships was run by Sir William Garthwaite. He was responsible for the final British classic sail training vessel, the four masted barque *Garthpool*, wrecked off the Cape Verde Islands in November, 1929. Sir William later founded the Sea Lion Training Ship Society which was dedicated to build or acquire an existing ship to continue that form of sail training. Unfortunately it came to nothing through lack of British public support, and classical sail training was doomed to come to a virtual finish with the end of the commercial sailing ship.

In Germany, however, the pattern was different. The need for officers to service the expansion in the steam fleet, coupled with the then common requirement that they be trained in sail, led to North German Lloyd building perhaps the most famous sail training vessel of all, the 3,242 ton, four masted barque, *Herzogin Cecilie*, in 1902 (see page 24). She was to earn her keep and her reputation in the South American nitrate trade, carrying 60–70 cadets at the same time. In 1921, as trade dwindled, she was bought with a number of other famous ships by Captain Gustav Erikson of Mariehamn. Erikson carried fee paying apprentices, but the only training provided was the experience of being part of the crew. Those aboard received no formal grounding in navigation or theoretical seamanship. This practice was part of the development that led to the understanding of the advantage of the adventure training vessel in that it combined a broadening of experience and the need to work with others in a different environment. A number of Erikson's trainees had signed on for just this and were not bent on working the promotion ladder of the merchant navy.

Remains of Fennia II in the Falkland Islands.

The Finns believed in the benefits of sail training. The Finnish School Ship Association owned the barques *Favell*, *Glenard* and *Fennia*. The Association had bought the four masted, 3,112 ton barque *Champigny* and renamed her *Fennia II* in 1925. With her extended poop, she made a fine cadet ship, but she had a short sea going life, ending her career, in common with others such as the Brunel steamer *Great Britain*, as a store hulk in the Falkland Islands. She is still there but, again like the *Great Britain*, there are plans for her restoration.

Almost all European countries, with the exception of Britain, maintained a tradition of training under sail and many of these subscribed to the sail training vessel concept of trading and training. The Belgians, for example, had the four masted barque *L'Avenir* (later *Admiral Karpfanger*, under the German flag), and the French, who had disposed of the similarly rigged *Champigny* (later *Fennia II*, see page 10), possessed for a short time the ex Flying P Line, four masted barque *Pola*, renamed *Richelieu*.

The end of sail training in this pure sense came with the retirement of the barque *Passat*, after the loss of the four masted barque *Pamir* (see page 32), in September, 1957, after being capsized by hurricane Carrie. Both ships were due for retirement because of competition from larger power driven bulk carriers, running then on cheap fuel.

With the last of the classic form of sail training vessels going into retirement, the training element had to be carried on aboard the cadet ship and it is to the credit of the German Navy that it commissioned the barque *Gorch Fock II* a year after the loss of the *Pamir*, believing that the sea and ocean environment had a lot to teach the Navy, even if sailors were professionally relieved of the day-to-day requirement of working on intimate terms with the elements.

The Development of the Cadet Ship

Most of the major naval powers in the 19th century gained sea experience either aboard regular men-of-war or attached vessels.

The British Royal Navy, for example, commissioned and lost the 921 ton, 26 gun Jackass frigate, *Eurydice*, of which Conan Doyle wrote:

> 'Up with her royals that top the white spread of her,
> Press her and dress her and drive her through foam,
> The Island's to port and the mainland ahead of her.
> Hoy for the Warner and Hayling and home.'

The poem goes on to describe how the 35 year old vessel was lost in a squall at the back of the Isle of Wight on 24th March, 1878.

1. H.M.S. Eurydice.

She had been commissioned in the year before her loss, as a training ship for the ordinary seamen of the Home Port Reserves. Her role was a clear one – to provide sea experience for those who were engaged in other pursuits, though at that time they may have been seamen, or longshoremen, in their own right. *H.M.S. Eurydice* differed from the Royal Naval training brigs of the 1890s in that they were attached to stationary training ships at Portsmouth, Devonport and Portland, for they were the training ships of the regular Navy. These little brigs had a reputation for smartness.

2. H.M.S. Martin.

In 1901, as Admiral Lord Cunningham, remembering his days in the brig *H.M.S. Martin*, recorded in his *Sailor's Odyssey*: 'Every boy joining the Navy was not only given sail drill in his training ship, but was sent for six weeks or more to a brig or sloop to gain experience at sea'. He further remarked that 'sail training certainly produced lads unsurpassed for smartness, alertness and physique'. The boys referred to were aged between 15 and 18 years.

The Naval training squadron of that time consisted of the 105 ft (32 m) brig *H.M.S. Martin* (ex *Mayflower*) of 508 tons displacement, attached to the stationary training ship *St. Vincent* at Portsmouth. The brigs *Pilot*, built in 1890, and *Nautilus* were the sea-going arm of *H.M.S. Impregnable* at Devonport, and the brigs *Sealark* and *Seaflower* served *H.M.S. Boscawen* at Portland. These little ships find their counterparts in the naval school ships of other nations today, such as the German Navy's *Gorch Fock II* (see page 126).

The Germans owe much of their interest in sail training to the Duke of Oldenburg and his German School Ship Association formed at the close of the last century. The Deutscher Schulschiff-Verein owned the *Grossherzogin Elizabeth*, now the French accommodation vessel *Duchesse Anne*; *Princess Eitel Friedrich*, now the Polish full-rigged ship *Dar Pomorza* (see page 136); the barque *Grossherzog Friedrich August*, now the Norwegian barque *Statsraad Lehmkuhl* (see page 64) and the full-rigger *Schulschiff Deutschland*.

All except the *Grossherzogin Elizabeth* were expropriated as part of the reparations payment after the First World War, and she went to France after the Second. The full-rigger *Schulschiff Deutschland* was not launched until 1927 and so Germany was short of cadet ships in the twenties. The government decided to buy the 373 ton Jackass barque *Niobe* (ex four masted schooner *Morten Jensen* built in 1913), and this reaffirmed the German belief in sail training, leading directly to the building of the Blohm & Voss barques. *Niobe* foundered after being capsized by a squall off the Fehmarn Light Vessel on 26th July, 1932, with a loss of 69 lives. However, the die was cast and *Gorch Fock I* (now the Russian barque *Tovarishch II*), the first of the German barques, was ordered by the government. This led to *Horst Wessel* (now the U.S. Coast Guard's *Eagle*, see page 106) in 1936 and *Albert Leo Schlageter* (now the Portuguese navy barque *Sagres II*, see page 118), and very nearly the same design was used for the Romanian *Mircea II* (see page 138) in 1938 and the present West German Navy barque *Gorch Fock II*, built in 1958. A second redistribution took place after the Second World War. It is interesting to reflect that past German naval sail training ships continue to play a major role in instructing the world's young naval officers.

The French acquired the full-rigger *Princess Eitel Friedrich* (now *Dar Pomorza*) after the First World War. She was handed over to the Société Anonyme de Navigation Les Navires Ecoles Français which owned the four masted barque *Richelieu*, a cargo/cadet vessel or classical sail training ship, and they renamed the German full-rigger *Colbert*, and then promptly laid her up.

Apart from the use by the naval college at Brest of the three masted Camper & Nicholson-built yacht *Ailee*, the French navy relied from 1932 to the present on the twin schooners *La Belle Poule* and *L'Etoile* (see page 114).

The famous American frigate *Constitution*, 305 ft (93 m) in length and 2,200 tons displacement, was used for the Naval College at Annapolis from 1865 to 1871. At the start of her career in this role, she was already 68 years old and had seen active service against the British, as well as taking part in cruises off the west coast of Africa to contain the slave trade. She was capable of some 13½ knots and carried 42,720 sq ft (4,433 sq m) of sail, including studding-sails, which compares with 45,000 sq ft (4,181 sq m) of the *Herzogin Cecilie* (see page 24). The *U.S.S. Constitution* and *Constellation*, which later also served as a naval training ship at Newport (Rhode Island) from 1894 to 1914, are maintained as museum ships, the former at Boston and the latter at Baltimore.

The U.S. Coast Guard, then the Revenue Service, started its cadets at sea aboard the topsail schooner *J. C. Dobbin* in 1877. She was followed by the 108 ft (33 m) barque *Salmon P. Chase* (see page 30), which had been built for the Service in Philadelphia in the same year.

Abraham Rydberg.

Training of this kind was not restricted to the large naval powers. Turkey, for example, complemented her naval school with a cadet ship in 1848. The brig *Nuveydi Futuh* was their second training ship.

However, training at sea in cadet ships was not just the prerogative of the navies, although the methods and smartness of the service were often adopted by the mercantile marine, as far as their training ships were concerned. A good example of this was the Swedish *Abraham Rydberg I*, owned by the Abraham Rydberg Foundation. She was built on the lines of a naval frigate and the Foundation saw to it that her handling matched her appearance. She was a cadet ship attached to the Foundation's shore establishment at Stockholm.

In contrast, and demonstrating the difference between cadet and sail training ships, one of her successors, the 270 ft (82 m), 2,345 ton, four masted barque *Abraham Rydberg III* (ex *Star of Greenland*) carried both cargo and cadets.

The Danes also adopted naval tradition aboard the 100 ft (30 m) full-rigged ship *Georg Stage I* (see page 16), built of iron in 1882. She was in the cadet ship tradition and had been endowed by the Danish shipowner Frederick Stage. *Georg Stage I* was run by the Stiftelsen Georg Stage Minde and carried a crew of ten and 80 cadets between the ages of 15 and 18 in the manner of *H.M.S. Martin*. The boys remained aboard *Georg Stage I* for the five summer months, while the Royal Navy cadets were only there for six to eight weeks.

The *Georg Stage I* was sold to Alan Villiers who, setting off in 1934, made an epic round the world voyage in her. Under his command, she was renamed *Joseph Conrad* and pioneered a new form of training under sail, the forerunner of adventure training schemes, which will be discussed later.

The ship is now owned by Mystic Seaport, Connecticut,

U.S.A. and she has returned to the role of accommodation vessel, as favoured by the navies of the last century. However, instead of working in harness with a sizeable sea training vessel, *Joseph Conrad* is now part of the 'Young Mariner' programme, using classrooms and dinghies for the benefit of school children and young people studying sailing, seamanship, navigation and maritime history. This is, of course, basically seamanship training allied to the cadet ship idea and now developed by such organisations as Britain's *T.S. Foudroyant* (ex *H.M.S. Trincomalee*), where courses are run for boys and girls aged 11 upwards, from March to September in Portsmouth Harbour. The 1,447 ton displacement *T.S. Foudroyant* was built of teak in Bombay in 1817 and is the oldest ship afloat today still engaged in a form of sea training.

The cadet ship allied to merchant navy training took many forms. *Lady Quirk*, tender to the Quirk Nautical Training College circa 1926, was one of the first in Britain. Another ship that helped to give cadets their first experience of a night at sea was the Southampton School of Navigation's ketch *Moyana* (see page 36). She was commissioned during the time of the Battle of the Atlantic in 1943, and had to restrict herself then to the enclosed waters of the Solent. For a considerable time, she was the only vessel of any size employed as a sail training vessel for the British merchant navy. She followed Devitt & Moore's last cadet ship, the three masted topsail schooner *St. George* mentioned earlier. *Moyana* was lost on passage home from Lisbon, after winning the first International Sail Training Race in 1956.

It is perhaps strange that the British, who had all but forsaken training under sail, should found the Sail Training Association that has organised the biennial races for training ships since 1956.

The Beginnings of Adventure Training Under Sail

The cadet ship concept, that is a ship devoted entirely to training with no other supporting role, developed in various ways as explored on page 42. This trend includes, of course, a strong emphasis on using the sea as a medium for adventure training. Most developed nations use the 'wilderness' land areas for adventure or character building purposes in one way or another. The sea and oceans are available too, and this resource has been exploited through the use of ships either specially built or adapted. The *Joseph Conrad* was one of the forerunners of this approach, although ships like the British schooner *Maisie Graham*, an ex German pilot schooner, were to develop the idea still further. *Maisie Graham* started her training career, following the cutter *Bonaventure*, as a school ship owned by the Graham Sea Training School at Scarborough, England. Like many others of its type, such as Tabor Academy, Marion, Massachusetts, U.S.A., this school provided a good education combined with nautical instruction. The schooner *Tabor Boy* remains an integral part of the Tabor Academy.

Maisie Graham made two cruises a year with her 12–16 year old crew until 1938, when she was sold to Gordonstoun School, near Elgin, Scotland, and her name was changed to *Prince Louis I*.

The founder of Gordonstoun, and Headmaster at that time, was Dr. Kurt Hahn, C.B.E., who was to play a central part in this form of educational development. The first master of the *Prince Louis I* after the change of ownership was Commander John Lewty, O.B.E., R.N., and Prince Philip, Duke of Edinburgh, was the first trainee.

In 1940 at the request of Dr. Hahn, Captain G. W. Wakeford, O.B.E., F.R.I.N. organised and ran what later became the Outward Bound Sea School at Aberdovey. When the course was developed and confirmed on a regular basis, the *Prince Louis I* was transferred from Gordonstoun to the new organisation, to join its first ketch, the ex French crabber *Garibaldi*.

1. H.R.H. Prince Philip aboard the Prince Louis II.

2. Prince Louis.

The selection of ships that follows provides an insight into the vessels that pioneered and developed the cadet and sail training ships and led, in turn, to the birth of the adventure training vessel.

MACQUARIE

(ex *Melbourne*, later *Fortuna*)

The ship-rigged *Macquarie* was built by R. H. Green of Blackwall, London, in 1875 as the *Melbourne*. This famous yard was responsible for a long line of wooden Blackwall frigates, but the *Melbourne* was built of iron, the third ship the yard had built of that material. It was originally intended to name the ship *Victoria*, and she carried a fine figurehead of the Queen. Her original role was as a fast passenger and cargo carrier on the Australian run, and R. H. Green, the builders, ran her as such until 1887.

In 1888, she was sold to Devitt & Moore – her name was changed to *Macquarie* and her port of destination to Sydney. Devitt & Moore had pioneered the idea of including a fair number of premium paying cadets with passengers, cargo and normal crew. The cadets would receive the usual education from a specially employed schoolmaster, as well as learning seamanship and navigation – the arts of the mariner – from the ship's officers (see introduction to this section).

In 1897, the *Macquarie* succeeded the ex Orient Line clipper *Hesperus*, which had, with *Harbinger*, successfully launched the scheme through Devitt & Moore's Ocean Training Ships, Ltd. Part of *Macquarie*'s passenger accommodation was given over to trainees, though she continued her run, carrying cargo, passengers and emigrants to Sydney. In tandem with *Illawarra*, she continued in this role until 1904, when they were both succeeded by the four masted barques, *Medway* and *Port Jackson*.

Johan Bryde, a Norwegian shipowner from Sandefjord, bought her, renamed her *Fortuna*, and converted her into a barque, using her for tramping until 1909. She was then sold to the Wallarah Coal Company in her old port of Sydney. She ended her life there, as a coal hulk, feeding steamers, until broken up in 1953.

LAST OWNERS: Wallarah Coal Company
PLACE OF BUILDING: Blackwall, London
RIG: Ship
DRAUGHT (feet): 23·7
BEAM (feet): 40·1
DESIGNER: R. H. Green
YEAR OF BUILDING: 1875
CONSTRUCTION: Iron
TONNAGE: 1,857 Reg. Tons
BUILDERS: R. H. Green
BUILT FOR: R. H. Green
L.O.A. (feet): 269·8
LAST NATIONALITY: Australian (as coal hulk *Fortuna*)

1. Under full sail: her figurehead is of Queen Victoria, after whom she was originally to be named.
2. The crew taking in the main course: she carried a fair number of premium paying cadets as well as passengers, cargo and her normal crew.
3. Making passage to Australia: until 1904, along with Illawara, *she transported emigrants from Britain.*

JOSEPH CONRAD

(ex *Georg Stage I*)

The full-rigged *Georg Stage I* (later *Joseph Conrad*) was designed and built of iron in 1882 by Burmeister and Wain, Copenhagen. She was endowed as a school ship and run by the Stiftelsen Georg Stage Minde. The Danish shipowner Frederik Stage had conceived this idea, following the death of his son Georg in 1880. The idea of naming a ship after a lost son was followed nearly 90 years later in 1967, when Sir James Miller presented a substantial sum toward the building of the Sail Training Association's second schooner, *Malcolm Miller*, in memory of his youngest son who had been killed in a car accident.

In her original guise, *Georg Stage I* was equipped with a two cylinder steam engine, boiler and lifting propeller, a common feature of the 'up funnel, down screw navy' of the day. Remains of this system still exist in the British Maritime Trust's frigate *Gannet* (later *Mercury*) and a pioneering equivalent is to be found in the propeller arrangement of Brunel's steamer *Great Britain* built in 1843 and now at Bristol.

The boiler and engine were open to the main cadet flat, where boys slept from hammocks slung from the deckhead. The hammocks themselves, when not in use, were stowed in the old Royal Navy manner in hammock nettings in compartments in the top of the bulwarks, and covered over with tarred canvas. The officers occupied four single cabins aft, with a central mess room and a separate saloon right in the stern. There were a couple of petty officers' cabins forward on the starboard side.

The figurehead was originally of wood but the trainees so disfigured it, that it was replaced in bronze. The *Georg Stage*'s rigging was that of a ship-rigged sloop of the Royal Navy in much the same way as the *Abraham Rydberg I*. Training was much on naval lines at this time.

Eighty boys aged between 15 and 18 and a permanent crew of ten served in the ship for five summer months. Her refitting was done partly by the boys themselves. They learnt about the ship, as well as how to work aloft and to handle sails while still at anchor or alongside, before going to sea in her. The engine was very strictly an auxiliary to the sails. The length of cruises was gradually worked up, until she crossed the North Sea to visit British ports as part of the schedule.

Disaster struck on the night of 25th June, 1905, when she was run down by the British steamer *Ancona* of Leith with the loss of 22 boys, many being drowned in their hammocks.

She was salvaged and rebuilt, the steam engine and boiler, which was regarded as little more than ballast, being removed, together with the lifting propeller. In order to ensure that she would not sink in so abrupt a manner again, water-tight bulkheads were installed.

The *Georg Stage I* remained without auxiliary power for the next ten years, when she was provided with a 52 hp semi-diesel for manoeuvring purposes.

Her work to improve the qualities of Danish seamanship was recognised and in 1934 it was decided to replace her with a new ship. *Georg Stage II*, about 20 ft (6 m) longer overall, was laid down and by chance, Alan Villiers heard of her predecessor when looking for a vessel 'to wander the tropic seas with a crew of lads'. He bought her, renamed her *Joseph Conrad*, and took her to Ipswich under the red ensign. There she was refitted to some extent by Whisstocks of Woodbridge. Alterations were made to the accommodation – cabins were provided for the master, mate, 2nd mate, spare for the pilot, five guests, four ABs, and a steward, and bunks were built in for 16 boys. The engine remained as ballast, a tradition now aboard this ship.

Arrangements were made to fit the vessel for ocean voyaging rather than spending the summer in the Baltic and North Seas. Her boats, normally slung outboard, were placed on skids and her various companions and openings made safe. Provision was made for more fresh-water tanks, and her ballast was increased.

On the 22nd October, 1934 she left Ipswich for New York, calling at Madeira and Nassau. She sailed under the burgee of the Royal Harwich Yacht Club and although she carried cadets, she was not registered as either a merchant ship or a sail training vessel by the Board of Trade, as it was difficult to fit her into either category under their rules.

She was nearly wrecked in New York after the anchor cable had parted. During a refit from this, her second near-disaster, the bronze figurehead of Georg Stage was removed to grace the new *Georg Stage II*, and one of Joseph Conrad was substituted.

The *Joseph Conrad* returned to New York on 16th October, 1936, after covering over 57,000 nautical miles. Under Alan Villiers, she had given her multi-national cadets a training as good as any provided by the school ship and had perhaps pioneered a new form of training which was more 'by the sea' rather than 'essentially for the sea'. This concept was to be exploited later in differing forms by adventure training vessels the world over.

Alan Villiers sold her to Huntington Hartford who refitted her as a yacht, giving her every modern convenience. During World War II Huntington Hartford presented her to the United States Maritime Commission based at St. Petersburg, Florida, which used her as a sea training vessel for the American Merchant Navy.

In 1947, by an Act of Congress, the *Joseph Conrad* was turned over to Mystic Seaport, a maritime museum in Mystic, Connecticut. She was towed from St. Petersburg to Mystic where subsequent surveys revealed that it would be uneconomical to modernise and refit her for further service at sea. In 1949, she was placed in service as a permanently moored berthing ship for the Museum's educational programmes, and over the ensuing years, thousands of school children, scouts and students have lived aboard her.

Mystic Seaport has taken steps to preserve the historic old vessel. In 1961 she was hauled out so that a coat of fibreglass could be sprayed on the hull to prevent corrosion. Her bilges have also been coated with ferro-cement. Other restoration projects will be completed well before the ship's hundredth anniversary in 1982.

To this day, the old ship imparts an atmosphere of tradition which is invaluable to the studies of the basic arts of sailing, seamanship, navigation and maritime history which take place aboard her.

JOSEPH CONRAD

OWNERS: Mystic Seaport
DESIGNER: Burmeister and Wain
BUILDERS: Burmeister and Wain
PLACE OF BUILDING: Copenhagen
YEAR OF BUILDING: 1882
BUILT FOR: Stiftelsen Georg Stage Minde
RIG: Ship
CONSTRUCTION: Iron
L.O.A. (feet): 100·8
DRAUGHT (feet): 13·2
BEAM (feet): 25·2
TONNAGE: 203 Reg. Tons
PRESENT NATIONALITY: American

1. A trainee at the helm.
2. Studding sails set : this fair-weather canvas gave a ship extra speed, more important in those days than economy of operation.

2

L'AVENIR

(Later *Admiral Karpfanger*)

The *L'Avenir* replaced the ship-rigged *Comte de Smet de Naeyer I*, the Greenock-built Belgian training ship, belonging to ASMAR (Association Maritime Belge S.A.). The *Comte de Smet de Naeyer I* which had been launched in 1904 was lost, with 33 lives, on her second voyage. The actual cause of her end while sailing down the English Channel on 18th April, 1906, has never been discovered.

ASMAR was formed in 1903 at Antwerp by Belgian shipowners and merchants to try to expand the number of deck officers required for their growing merchant fleet. The scheme was subsidised by the Belgian government until the government decided to run its own training scheme and built the barquentine *Mercator* in 1932. The Association had also bought the 1,357 ton, Port Glasgow-built *Linlithgowshire* (ex *Jeanie Landels*, launched in 1877) to fill the gap caused by the loss of the *Comte de Smet de Naeyer I*. The second *Comte de Smet de Naeyer* was responsible for the pre-training of cadets and had accommodation for around 100. She continued in this role until the scheme was finally wound up in the early 1930s.

When launched in 1908, the *L'Avenir* had the same appearance as the larger *Herzogin Cecilie* (see page 24). This may not have been too surprising, for Rickmers built both ships within six years for the same purpose, that of training while trading. *L'Avenir* had a long poop and a deck-house stretching aft into the well deck from the small to'gallant fo'c'sle. There was a fore and aft bridge connecting the fo'c'sle with the poop, by way of the deck-house.

The *L'Avenir* was one of the first vessels to have wireless, and carried her aerials on two distinctive poles on the fore and mizzen masts. The cadets' accommodation, for about 60, was in the centre section of the poop, with classrooms forward of their hammock space.

In the First World War, which broke out at the end of her sixth voyage, she dispensed with training but continued to carry cargo, dropping her white training ship colours in favour of the more discreet black with buff masts and spars. She managed to survive the war without major incident and then returned to her original job of training. It was soon found, however, that her accommodation was insufficient, and as a result, in 1921, the to'gallant fo'c'sle was extended aft to include the whole of the deck-house. This reduced the well deck to barely 36 ft (11 m). The ship's boats were rearranged at the same time, the lifeboats and Captain's gig being placed on skids just aft of the midship wheel platform. The two quarter boats remained on davits, as originally placed.

L'Avenir now had accommodation for 80 cadets and proved a most successful vessel as far as ASMAR was concerned. In 1926 she won the grain race, Geelong to the Lizard, in 110 days. As mentioned earlier, it was only the Belgian government's decision to remove her training subsidy in favour of starting its own scheme with *Mercator*, that persuaded the Association to put the barque up for sale in 1932.

The British Sea Lion Sail Training Ship Society, founded and chaired by Sir William Garthwaite in 1930 (see Foreword to this section) was interested in her. Sir William was the owner of the late four masted barque, *Garthpool*, wrecked in November, 1929, in Boavista Bay, Cape Verde Islands, and was a strong believer in the value of sail training in large, commercial vessels. However the Society was unable to raise the necessary finance or to persuade either the British Royal Navy or the merchant service to accept the idea of sail training in anything much bigger than a whaler.

Gustav Erikson of Mariehamn, however, knew of *L'Avenir*'s reputation and how to make a success of sail, for he had by this time a sizeable fleet of his own.

The Mariehamn shipowner altered her to provide passenger accommodation for first and third class passengers. The voyage out to Australia and back worked out at 50p per day for berth and food, but the more popular trip was the short haul back to Mariehamn from Glasgow after discharging grain, or from Mariehamn to Copenhagen on the outward voyage. These trips were such a success that Erikson had a summer schedule of passenger trips from Mariehamn into the Baltic, using an accompanying tug to ensure that the timetable was maintained.

However, such innovation, even though a success, was not justification alone for keeping *L'Avenir*. Her grain capacity was not great enough to ensure a proper trading return and so Erikson sold her in 1937 to the Hamburg America line. Nord Deutscher Lloyd had also looked at her as a replacement for the *Herzogin Cecilie*. She was altered, in fact, to a cadet ship for her new owners. Her name was changed at the same time to *Admiral Karpfanger* and under that name she left Hamburg for Australia on 18th September, 1937, arriving at Wallaroo on 5th January, 1938, in a time of 107 days.

After loading grain at Port Germein, she left for Hamburg, and on 1st March gave her position as 51° S 172° E. She was never heard of again, though small pieces of wreckage identified as belonging to her were found on the shores of Tierra del Fuego.

There is no doubt that *L'Avenir* was one of the most successful of the trading and training vessels.

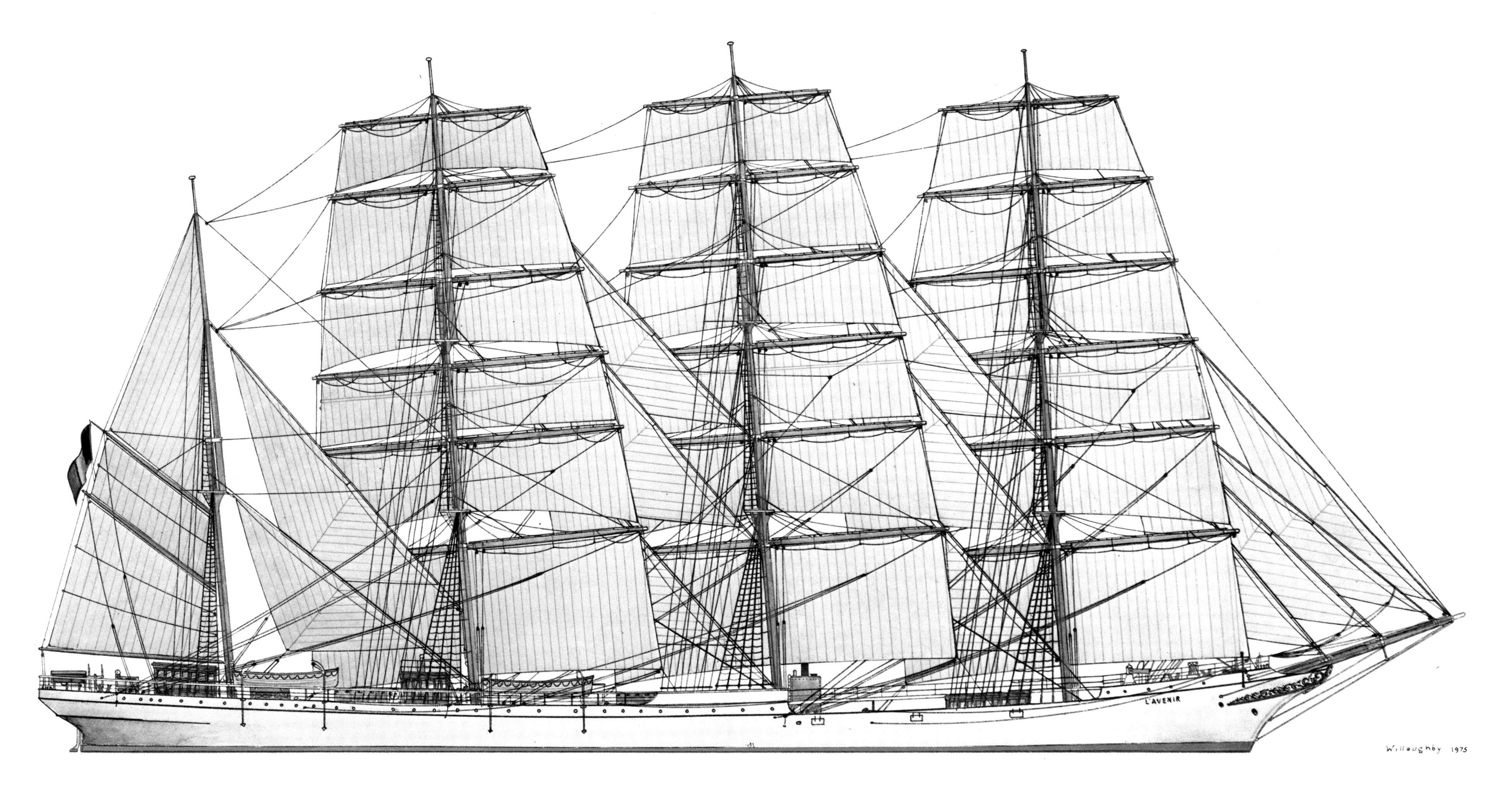

L'AVENIR

LAST OWNERS: Nord Deutscher Lloyd
PLACE OF BUILDING: Bremerhaven
RIG: 4-masted barque
DRAUGHT (feet): 26·5
BEAM (feet): 44·8

DESIGNER: Rickmers of Bremerhaven
YEAR OF BUILDING: 1908
CONSTRUCTION: Steel
TONNAGE: 2,738 Disp.

BUILDERS: Rickmers of Bremerhaven
BUILT FOR: Association Maritime Belge
L.O.A. (feet): 278·2
LAST NATIONALITY: German (as *Admiral Karpfanger*)

1. Down-channel with the White Cliffs of Dover to starboard: in 1926 she won the grain race Geelong – the Lizard in 110 days.
2. View of the compass platform from the mizzen top.
3. Chipping and painting the hull, a harbour job: in wartime her white livery was changed to a more discreet black with buff spars.
4. Under full sail with triangular main and cro'jack.

HERZOGIN CECILIE

Herzogin Cecilie, Nord Deutscher Lloyd's famous training ship, owed her existence to the growth of the steamer. The German merchant navy demanded that candidates for a master's ticket should have a sufficient background of sail training and to keep up with the training and trading programme, the navy bought the Rickmers' iron barque, *Albert Rickmers* (2,395 tons gross), that had been launched in 1895. She was renamed *Herzogin* (Duchess) *Sophie Charlotte* as a compliment to the then President of the German Training Scheme, The Grand Duke of Oldenburg. This barque was a fast and successful ship, holding the record of 21 days from Hiogo to Portland, Oregon, and managed Sydney to Europe in 77 days.

On 23rd April, 1902, as testimony of the success of the training/trading scheme, Nord Deutscher Lloyd launched the four masted barque *Herzogin Cecilie* (3,242 tons gross), named after the Duchess Cecilie of Mecklenburg, who later married the son of Kaiser Wilhelm II. She had a large cargo capacity and better accommodation than the *Herzogin Sophie Charlotte* and cost her owners £43,000, being built regardless of cost as a cadet ship, to earn her living in the nitrate trade from the west coast of South America. She was constructed of steel with a double bottom, with three watertight bulkheads and two decks, again of steel sheathed in wood. Her combined poop and midship house ran 194 ft (59 m), over half her length. This was a great advance over the orthodox sailing vessel, where the main deck was often drowned in water which drained away painfully slowly through the wash ports. This caused the accommodation to be almost permanently damp and hot food a luxury. The helmsman, too, was reasonably secure amidships, another unusual feature. He stood high with the shelter of a large chart-house behind him.

The *Herzogin Cecilie*'s normal complement was five officers, a doctor and two schoolmasters, 14 petty officers, ordinary seamen, cooks and stewards, and 60–70 cadets, a total of around 90 aboard. The cadets paid £40 all found, including schooling, and after three years and an examination passed out as fourth officers.

On her maiden voyage from Bremen to Astoria, via the Horn, *Herzogin Cecilie* experienced perhaps her most testing time. She was struck by a Pampero, a sudden squall born of the pampas, and lost her spars, making a call at Montevideo necessary for repairs.

In the years leading to the First World War, the barque was commanded by three masters – Captain Deitrich until 1909, under whose command she visited Japan and circumnavigated the world, Captain Walter to 1914 and Captain Ballehr until she was interned for the duration, months later. She and the similar *Herzogin Sophie Charlotte* were held for six years at Guayacan, until released to suffer confiscation under the Reparations Treaty.

Her second life began in the December of 1921, when she was bought by Captain Gustav Erikson of Mariehamn, Åland. He paid the bargain price of £4,250 for her and put her under the command of Captain de Cloux.

Erikson limited her employment to carrying grain cargos, and cut down her crew, as well as dispensing with the North German cadet system in favour of apprentices. These were boys of over 16 years of age, who paid for their first and second voyages and were then rewarded with the rating of ordinary seamen. Not all apprentices were making the sea a career and so in a sense the *Herzogin Cecilie* was one of the forerunners of today's adventure training vessels. Certainly this form of 'university' created fine men and excellent seamen. Alan Villiers once remarked that boys of 19 have no fear and can become proficient before self-preservation numbs them. Erikson was a master of this principle, and the *Herzogin Cecilie* was one of his finest tools.

Captain de Cloux did not drive her, with her smaller crew, as she had been pressed by her North German masters, but he made some fast passages. In the eleven seasons 1926–1936, she won the grain race four times, in 1927, 1928, 1931 and 1936. However, she had her fair share of trouble. In 1928 shifting cargo nearly capsized her off the Orkney island of Ronaldsay and in 1935, while in Belfast, the donkey boiler exploded, causing damage but no loss of life. Later that year she ran down the trawler *Rastede* off the Baltic island of Anholt.

But the final tragedy was yet to come. After a record breaking voyage from Australia, she arrived at Falmouth for orders on 23rd April, 1936.

On receiving orders to discharge grain at Ipswich, *Herzogin Cecilie* left Falmouth on Friday, 24th April. That night in thick fog, with a heavy swell but a light wind and, therefore, full sail, the barque struck the Ham Stone Rock, west of Salcombe, while making 7 knots. She was holed in the forehold and carried inshore by the swell, grounding in Sewer Mill Cove. The Coast Guard had noticed her distress rockets and the Salcombe Lifeboat was quickly on the scene.

The situation looked hopeless to the French salvage tug *Abeille No. 24* and work was started to remove all salvable gear as well as part of the cargo of grain that was untouched by sea water. About 500 tons were saved out of the 4,240 shipped.

Lady Houston quickly offered to pay the cost of salvage and repair and give her to the Royal Navy as a training ship, but this was refused. Captain Sven Erikson appealed to the public for assistance, saying that he would carry British cadets free of charge. The money rolled in, the ship was refloated – it looked as if her luck had turned. However, during the temporary repairs and removal of the fermenting cargo in nearby Starehole Bay on the 17th July, 1936, a severe gale sprang up, scouring the sand from under her bow and stern, leaving her supported amidships. Accounts talk of rivets popping as she broke her back and the brave cause of salving her became hopeless.

The 50 miles eastward out of Falmouth had proved the end of her. Her Master told the Receiver of Wrecks that the loss was caused by a combination of fog and a magnetic anomaly. Relics of her are now at Gustav Erikson's Maritime Museum at Mariehamn. The hulk was sold in the end to a local scrap merchant for £225.

HERZOGIN CECILIE

LAST OWNER: Captain Gustav Erikson
DESIGNER: Rickmers of Bremerhaven
BUILDERS: Rickmers of Bremerhaven
PLACE OF BUILDING: Reismühlen
YEAR OF BUILDING: 1902
BUILT FOR: North German Lloyd
RIG: 4-masted barque
CONSTRUCTION: Steel
L.O.A. (feet): 314·1
DRAUGHT (feet): 23·8
BEAM (feet): 46·0
TONNAGE: 3,242 Disp.
LAST NATIONALITY: Finnish

Getting under way.

1. Working on the mizzen upper to'gallant in heavy weather: on 31st March 1909, while making for the Horn from Port Augusta, she was pooped by a freak wave that rushed forward, sweeping away a lifeboat, the aft pair of wheels and long skylight, and demolishing the charthouse.
2. Hauling in on the headsail sheets in dirty weather.
3. Looking aft from the bowsprit.

1. Homeward bound: Herzogin Cecilie *was a smart ship and her speed was as impressive as her appearance. She averaged Europe – Chile between 1904 and 1914 in 74 days, with a best of 68; and Chile – Europe between 1905 and 1914 in 76 days, also with a best of 68 days.*

2. Awash in Sewer Mill Cove, west of Bolt Head, England, on 25th April, 1936: in fog at 4 am, with a heavy swell, she struck the Ham Stone. In spite of herculean efforts, and near success, this was the end of the finest training barque to have sailed.

3. Captain's inspection: Captain Erikson and Mrs. Erikson were on their honeymoon when the ship struck.

SALMON P. CHASE

The American Revenue Service, later the United States Coast Guard, started training its cadets at sea with the topsail schooner *J. C. Dobbin* in May, 1877. The Coast Guard entrusted the ship and the programme to Captain J. A. Henriques, two lieutenants, and, to use the contemporary description, a few 'intelligent petty officers'.

The ship and the scheme were a success and as a result, the *Dobbin* was replaced by the 250 ton, 106 ft (32 m) barque, *Salmon P. Chase*, which had been specially built in the yard of Thomas Brown & Sons of Philadelphia, in 1877. Like the first and second *Abraham Rydberg* of Sweden and the first and second *Georg Stage* of Denmark, the barque was a small edition of a larger ship of the period, but while the Swedish and Danish ships were designed for naval requirements, the *Salmon P. Chase* was designed for commerce. Her fine run, clipper bow and rounded stern gave her a 'tall ship' look. She operated from New Bedford, and in the early part of her career, she followed the pattern established by her predecessor, the *J. C. Dobbin*, of sailing during five months of the summer along the coast between Buzzards Bay, Massachusetts and Chesapeake Bay, Maryland. She also made passage between the United States and Bermuda, and this was later extended to include a European cruise. The *Chase* had a good reputation for both speed and seaworthiness.

During the first part of her career, the cadets had a winter school in an old sail loft at New Bedford. Later the *Chase* continued training during the winter with classrooms aboard, moving to southern ports to avoid the harsh weather. Winter quarters for the Coast Guard were established ashore at Arundel Cove, Baltimore, Maryland in 1900.

Academy training was suspended between the years 1890 and 1894, and soon after this the barque was enlarged – she was cut in two and a new midships section fitted in order to increase and improve the accommodation.

She was in collision off Charleston soon after this, but was repaired and remained in service with the Coast Guard until 1907. By this date, she had been in service for nearly 30 years and it was decided to pension her off to become an accommodation vessel at the Coast Guard Academy's winter quarters at Arundel Cove. She remained in this role until rammed and sunk in 1930.

The Coast Guard replaced *Chase* with the ex Navy practice ship *Bancroft*, renamed *Itasca*. Her successor, the gunboat *Alexander Hamilton*, named after the founder of the Coast Guard, remained in service until 1930, by which time the Academy had moved to New London.

Sail training was carried on in the *Gloucester* fishing schooner, renamed *Chase*, the 65 ft (20 m) yacht *Curlew* and the ocean racing schooner *Atlantic*, which may be salvaged and restored, after spending 20 years at Wildwood, New Jersey, 10 of these on the bottom.

In 1942, the 700 ton ship-rigged *Danmark* (see page 52), was placed at the Coast Guard's disposal, and she was replaced in 1946 by the German 1,800 ton barque *Horst Wessel*, which the Coast Guard renamed *Eagle*.

LAST OWNERS: U.S. Coast Guard Academy
DESIGNER: Capt. Merryman
BUILDERS: Thomas Brown & Sons, completed by Joseph Allen
PLACE OF BUILDING: Philadelphia
YEAR OF BUILDING: 1877
BUILT FOR: United States Coast Guard Academy
RIG: 3-masted barque
CONSTRUCTION: Wood
L.O.A. (feet): 106·0
DRAUGHT (feet): 11·0
BEAM (feet): 25·5
TONNAGE: 250 Disp.
LAST NATIONALITY: American

1. Chartwork: Chase *operated summer training cruises out of New Bedford.*
2. In light weather, with Revenue Service flag on the fore.

2.

PAMIR

The four masted steel barque, *Pamir*, was built in 1905 by Blohm & Voss at Hamburg for Ferdinand Laeisz's Flying P Line. She was designed for the Chilean nitrate trade and was strongly built, having a central bridge deck in common with the other ships of the line.

The *Pamir* was still owned by the Flying P Line and working at the nitrate trade at the outbreak of the First World War. She was then on her return to Europe with a full cargo of nitrate, and sought refuge in the Canary Islands where she was interned and unable to discharge her cargo until the war was over. She was then handed to the Italians as part of the reparations payments and they kept her penned in the Mediterranean. Laeisz had lost all 14 ships of his fleet in similar manner, but between the wars he managed to buy back six of them, including *Pamir* in 1925. He also built the larger *Padua* and *Priwall* to supplement their operations.

However, hauling nitrate by sail had ceased to be economical by the early 1930s and Laeisz gradually sold all his sailing ships, except the most recent. *Pamir* followed *Penang*, *Pommern*, *Ponape* and *Passat* to Captain Gustav Erikson, the remarkable Åland shipowner. Erikson employed *Pamir* mainly on the grain trade. In 1939, after discharging grain at Southampton, 96 days out of Australia, *Pamir* made for Gothenburg. She was laid up for the first year of the European war, before returning to her old guano trade, loading at Mahé for New Zealand and then making another trip to the Seychelles from Wellington. However, by then circumstances had changed. During her first voyage Finland was neutral but by the time *Pamir* reached Wellington, Finland had become an enemy.

The barque was seized as a prize and handed over to the Union Steamship Company of New Zealand, which completely refitted her and then operated her successfully between New Zealand and San Francisco, carrying wool from the former and grain from the latter. In the years 1943–1949, she made a return of some £30,000.

However, as the operation began to become unprofitable, it was decided to sell her. The New Zealand government was pressed to retain her as a sail training ship, but it was felt that the country's merchant navy did not merit such a vessel and *Pamir* was therefore laid up in 1949 at Penarth, after discharging grain from Port Victoria. She was joined in Penarth by *Passat*. The future looked bleak for these two large survivors of the grain fleet, and early in 1951 both were sold to an Antwerp shipbreaker. However a reprieve came at the eleventh hour, when they were purchased by Heinz Schliewen of Lübeck, who was convinced of a future for training while trading under sail.

Pamir and *Passat* were refitted as cargo/cadet ships and each was given an ex U-boat diesel engine as an auxiliary motor. Accommodation was provided for approximately 70 cadets. But costs were against the scheme and after one voyage by both vessels to South America, both ships were laid up, *Pamir* at Travemünde.

In 1954 *Pamir* and *Passat* were sold to the Landesbank Schleswig Holstein, acting for a consortium of some 40 shipowners, who had combined for the benefit of training their cadets. Both vessels were refitted, this time to carry 50 cadets each. However, by 1957, powered bulk carriers robbed the venture of its economic advantage and both ships were due to be laid up.

Tragically, in September of that year, *Pamir* on passage home with a cargo of 4,000 tons of Argentine barley was capsized by hurricane Carrie and lost south-west of the Azores, an area in which winds of such strength seldom occur. Eighty-six of her crew were lost, of whom 54 were cadets. There were two cadets among the six survivors picked up from a lifeboat by the American ship *Saxon*. Five other occupants of the lifeboat died in the 54 hours that elapsed before they were rescued following the distress call that had been received by the British cargo vessel *Manchester Trader*.

Passat had experienced similar weather but escaped with a certain amount of damage. Tests were carried out on her to find out how such a disaster could have overtaken the *Pamir*, for she had in the past been through conditions as bad, if not worse.

The inquiry suggested that her loss could have been caused by a combination of shifting bulk grain, incorrectly stowed, and too much sail. It was usual to stow grain in bags, slitting some of them to fill the spaces in between, thus creating a solid mass. *Pamir* had been carrying foresail, topsails and staysails in hurricane conditions, without compensating use of water ballast, for her tanks designed for this purpose were filled with grain.

Pamir and *Passat* were the last of the pure sail training vessels – that is, they followed the tradition of training and trading under sail. The economics of the trading operation brought their remarkable training qualities to an end and *Pamir*'s loss acted as a major deterrent to those who would try again, until the German Navy, with great courage and sea sense, commissioned the 1,727 ton barque *Gorch Fock II* (see page 126) in 1958. This move confirmed the value of a form of training that might have otherwise been lost to Germany, if not elsewhere.

LAST OWNERS: Landesbank Schleswig Holstein Consortium
DESIGNER: Blohm & Voss
BUILDERS: Blohm & Voss
PLACE OF BUILDING: Hamburg
YEAR OF BUILDING: 1905
BUILT FOR: Ferdinand Laeisz
RIG: 4-masted barque
CONSTRUCTION: Steel
L.O.A. (feet): 316·0
DRAUGHT (feet): 26·2
BEAM (feet): 46·0
TONNAGE: 3,020 Disp.
LAST NATIONALITY: West German

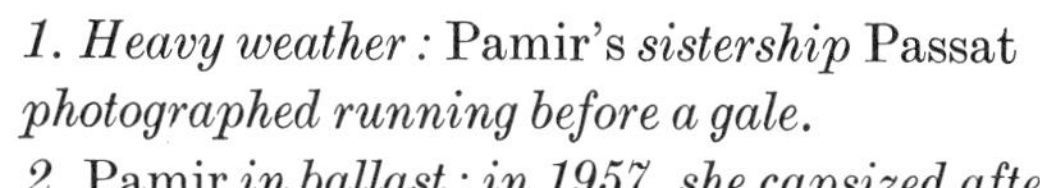

1. *Heavy weather:* Pamir's *sistership* Passat *photographed running before a gale.*
2. Pamir *in ballast: in 1957, she capsized after being struck by hurricane Carrie.*
3. *The sailmaker at work: chafe was the great enemy.*

ABRAHAM RYDBERG I

The *Abraham Rydberg I* succeeded the brig *Carl Johnan* as the second training vessel to be owned by the Abraham Rydberg foundation. Abraham Rydberg discovered the value of well trained men as a ship owner and, therefore, determined to leave enough to found a navigation school in Stockholm. He died in 1845 and the school soon prospered. By 1878 its success demanded the building of a new and larger vessel. The *Carl Johnan* was sold and the new wooden vessel laid down at Andersson's yard, Karlshamn.

In *Sail Training and Cadet Ships*, Underhill remarks 'that she was the smallest full-rigged ship ever built, with the appearance of a frigate in miniature, carrying single tops'ls, single to'gallants and royals, while forward she had a standing bowsprit and a long jib boom of the naval ships of her day'.

In 1911, she was replaced by *Abraham Rydberg II* (later the *Seven Seas*), a 262 ft (80 m) steel, full-rigged ship, which was succeeded in turn by the 4 masted cargo carrying cadet barque *Abraham Rydberg III* (ex *Star of Greenland*).

Abraham Rydberg I and, indeed, *II* were employed as the sea going part of the school. The arrival of a third ship to bear the name marked a change in policy; from then onward the training was to be at sea only, while the ship went about her business. In fact, the Foundation was adopting the classical form of sail training, as discussed in the preface to this section.

Abraham Rydberg I, after 32 years service to the Foundation, became the plain *Abraham* and reverted to commerce as a barque, gradually degenerating in traditional Baltic fashion, to a three masted topsail schooner. She ended her sea career as part of a ship yard jetty after some 50 years' service.

LAST OWNERS: Hälsö Shipyard
PLACE OF BUILDING: Karlshamn, Sweden
RIG: Ship
DRAUGHT (feet): 11·5
BEAM (feet): 22·3
DESIGNER: D. R. Andersson
YEAR OF BUILDING: 1879
CONSTRUCTION: Wood
TONNAGE: 149·0 Reg. Tons
BUILDERS: Andersson's Yard
BUILT FOR: Abraham Rydberg Foundation
L.O.A. (feet): 101·0
LAST NATIONALITY: Swedish (as *Abraham*)

A late picture of Abraham Rydberg II *in white training ship colours. She was one of the best-looking full-rigged ships ever built.*

MOYANA

The ketch *Moyana* was designed by Frederick Shepherd and built as the yacht *Nereus* in 1899 by White Brothers of Southampton. She was owned and raced in the short period 1928–1929 by the Earl of Dunraven, whose father had unsuccessfully challenged for the America's Cup with *Valkyrie II* and *III* in 1893 and 1895.

In the autumn of 1942, the founder and principal of the Southampton School of Navigation, Captain G. W. Wakeford, O.B.E., decided to look for a sea going training vessel. Training at sea for navigation schools at that time was difficult, with war time restrictions making cruises of any length out of the question. The vessel, therefore, needed to content herself and her trainees with the enclosed waters of the Solent, between the Isle of Wight and the mainland.

Captain Wakeford and his governors considered three types of vessel – a steam or motor vessel, a small trading schooner or brig with auxiliary engine, or a large, fast cruising yacht, again with an auxiliary engine. They decided against the power-driven vessel, because apart from those engaged in coastal navigation and the cadet at the helm, there would be little to do. They were also against the small trading vessel because of its lack of speed and manoeuvrability. Sailing vessels, though, had the added advantage of sailors being involved with the maintenance of spars and the overhaul of standing and running rigging in the annual refit. In the words of those responsible for the search for a training ship, 'The hazard of working a sailing vessel provided a sense of responsibility, alertness, awareness of danger and physical hardiness, all of which are of vital importance to sailors'. The 95 ft (29 m) ketch *Moyana* was purchased in 1943. She was large enough and carried 3,780 sq ft (351 sq m) of sail, thus avoiding the criticisms of small sailing boats and power-driven craft, which had arisen during the search. These criticisms were that small sailing boats were not massive enough to instil into the cadet the need for caution, and that power-driven craft provided little real training for an embryo merchant seaman, except for steering or washing paint. Neither brought him into any real contact with the sea, as does the sailing craft, which is directly governed by cadets who run the ship.

Moyana suited the school well and proved herself a success. After the war she extended her range to cover the north coast of France, the southern North Sea, and visits to the Port of London. The Southampton School of Navigation was, for a considerable time, the only such establishment in the British Isles employing a sail training vessel.

In 1956 *Moyana* competed in the first International Sail Training Race from Torbay to Lisbon. She completed the course in six days, eight hours and three minutes, arriving Lisbon at 2200 hours on the 13th July and so winning on handicap from the Norwegian ship *Christian Radich* (see page 66), by one hour and three minutes. During the race the wind never exceeded moderate to fresh in strength, giving her speeds for long periods of between 9½ and 10 knots.

Moyana left Lisbon on the morning of Thursday, 19th July, in a fresh northerly wind, which continued to increase in strength to gale force late the following day. Captain H. Stewart, her master, and later principal of the school, decided to heave to, to avoid strain and damage to his vessel. It was by then blowing up to 40 knots and a temporary watch keeper, Commander A. J. Stewart, R.N., was washed overboard but fortunately later recovered.

The gale continued for three days and by this time it was necessary to pump every watch. To add to the difficulties, one of the cadets developed suspected appendicitis. However, the wind began to drop away and by Saturday, 28th July, *Moyana* was 50 miles WSW of the Scilly Isles. The brief respite did not last, however, and conditions again deteriorated in the afternoon of Saturday, 28th July. *Moyana* again hove to and by this time was experiencing hurricane conditions. By 0215 hours on Sunday morning, it was decided to ask for help. The radio message was received by the *Robert Dundas* and relayed to Land's End radio, which broadcast the distress call to all ships.

This summons was received by the *Clan MacLean*, commanded by Captain Cater. *Moyana* was first sighted by an R.A.F. Shackleton which directed the rescue ships. Captain Cater managed to provide a lee for *Moyana*, by this time severely strained and making water.

All the crew were transferred to *Clan MacLean* without accident to officers or cadets, due to the skill of the master of the *Clan MacLean* and the seamanship of Captain Stewart of *Moyana*. *H.M.S. Orwell* subsequently tried to take *Moyana* in tow. A boarding party was successfully put aboard but to no avail. The men were rescued just before the ship sank later on Sunday, 29th July, 1956.

However, her purchase in 1943 and her foundering after victory contributed to the inspiration for the building of the Sail Training Association's two schooners, the *Sir Winston Churchill* and her sister *Malcolm Miller* (see page 90). Captain Wakeford had been convinced of the worth of sail training and immediately set about designing a larger replacement for the *Moyana*, should the money become available. The plans for this vessel provided the basis for the Association's first training ship, the *Sir Winston Churchill*.

LAST OWNERS: Southampton School of Navigation
DESIGNER: Frederick Shepherd
BUILDERS: White Bros.
PLACE OF BUILDING: Southampton
YEAR OF BUILDING: 1899
BUILT FOR: As a yacht for C. S. Guthrie
RIG: Ketch, built as yawl
CONSTRUCTION: Wood
L.O.A. (feet): 80·5
DRAUGHT (feet): 11·4
BEAM (feet): 17·5
TONNAGE: 103 T.M.
LAST NATIONALITY: British

1. Deck work: life aboard a sailing vessel brings the cadet into real contact with the sea.
2. Creole *and* Moyana *at the start of the first Tall Ships Race in 1956.*

2

PORT JACKSON

The *Port Jackson*, which came late into sail training, is generally considered one of the most beautiful iron ships ever built. It was a matter of company pride in the early 80s, that a new ship should be superior to her sisters in design, rig, strength and seaworthiness.

Duthie Bros., the owners, were ship builders themselves, but gave the contract to build to Halls. She was designed by Alexander Duthie for the Australian wool trade, operating out of Port Jackson, from where she took her name. Captain Crombie was her first master and her passages varied between 80 and 85 days outward and 90 to 95 on the slower passage back to Britain. Her best run in 24 hours was 345 miles, an average of over 14 knots.

A cutting in the company's voyage book records a voyage of 1887, when she was still owned by Duthie's and commanded by Captain Crombie.

'Melbourne, Nov. 5th – Port Jackson ship, at Sydney from London, reports that on September 12th she had to heave to in a fierce NNW gale, and while heaving to, one of the apprentices fell overboard and was drowned. The fore and mizzen topsails and several other sails were blown away decks swept of all gear, and the cargo shifted. In Lat. 39 S long 4 W, she was again hove to and again lost several sails; and on 18th ult. in Lat. 43 S long 132 E a sea broke over her at the fore rigging and smashed two of the boats, stove in the deck-house, started some stanchions and made a clean sweep of everything moveable on deck.'

Clement Jones in a short history of Devitt & Moore, *Sea Trading & Sea Training*, emphasises this as commonplace, saying 'there is very little incident to record in these steady passages to and from Australia until 1893'. He goes on to record the fire in Australia of 25th May of that year. The voyage book cuttings tell the story graphically. '*Port Jackson*, scorched throughout; has sustained extensive damage, mostly amidships; 13 hold beams very badly damaged, about 100′ lower deck burnt, main deck plates bent and wooden deck started, must be replaced, must remove portion cement. Costs through repairs estimated (?) £4,100 by surveyors inclusive docking, exclusive sails, ropes, provisions.'

However, she was repaired and left Newcastle, New South Wales on the 29th November, 1893, reaching London in 107 days.

Captain Hodge had command through the 90s and the *Port Jackson* managed to successfully compete against steam until 1904 when she was laid up in the Thames where she remained until 1906.

It was in this year that the ship took on a new role and the one for which she would be popularly remembered. For in 1906 Devitt & Moore had contracted to take a hundred *Warspite* boys of the Marine Society of London, on a round trip to Australia. They had no suitable ship available and so bought the *Port Jackson* and refitted her for training.

The first voyage was nearly a disaster, for having lost the newly appointed Captain Cutler who died and was succeeded by Captain Ward, she damaged herself leaving the West India Dock in London. But this was not the end of problems in her new role that year, for she was run down by the German steamer *Pyrgos* in the Downs, while at anchor in thick fog. However, she was soon repaired and made Australia and back, proving the training scheme a success. In 1907 this led to the *Port Jackson* being included in Devitt & Moore's cadet training service under Captain Maitland, taking 24 cadets and 50 boys from the *Warspite*.

In 1909 the service was taken over by a new arm of the Company, Messrs. Devitt & Moore's Ocean Training Ships, Ltd., in which other companies, such as Cunard, took shares, to look after their own training requirements.

In 1910 she was joined in the new Company by the four masted barque *Medway* and by carrying cadets and wool on the round trip to Australia, the training scheme flourished until the outbreak of the First World War.

Ships were then much in demand and the barque was off to Buenos Aires, later calling at New York where the cadets again joined her, outward bound for Adelaide and back to Nantes. After this successful, though problematical voyage, she was sold to the Swift Steamship Company of West Hartlepool and Captain Maitland retired.

Captain Maitland's mate took over, sailing in ballast again for Buenos Aires, where she loaded wheat for Britain. She left there on 17th January, 1917 and was torpedoed by a German U-boat off the Irish Coast on 28th April. The master and twelve men were lost, the rest taking to the boats and making Queenstown, Ireland (Cobh).

The Company history records how the old bosun was among those lost. He had retired with Captain Maitland, but returned to sea because of the darkened streets caused by Zeppelin raids. He told the Company that 'he felt safer at sea'.

PORT JACKSON

LAST OWNERS: Swift Steamship Company
PLACE OF BUILDING: Aberdeen
RIG: 4-masted barque
DRAUGHT (feet): 25·2
BEAM (feet): 41·1

DESIGNER: Alexander Duthie
YEAR OF BUILDING: 1882
CONSTRUCTION: Iron
TONNAGE: 2,212 Disp.

BUILDERS: Alexander Hall & Co.
BUILT FOR: Duthie Bros.
L.O.A. (feet): 286·2
LAST NATIONALITY: British

2

1. Looking forward from the poop: Port Jackson's *best run in 24 hours was 345 nautical miles, an average of over 14 knots.*
2. Captain Maitland and his Officers: Captain Maitland took over in 1907 when Port Jackson, *recently acquired by Devitt and Moore, started taking* Warspite *boys, under their new training scheme, on return voyages to Australia.*
3. Making sail: the main lower to'gallant being broken out. Port Jackson *was considered one of the most beautiful iron ships ever built.*

3

Ships of the present

'A year or a year and a half of training in a sea-going ship I would regard for a boy destined for the sea as a course in classical practice of the sea. What he will actually learn on board that ship he will leave behind him directly he steps on the deck of a modern steamship, but he will have acquired the old lore of the sea which has fashioned so many generations down to his very fathers, and which, in its essence, will remain with the future generations of seamen, even after the day when the last sail and the last oar have vanished from the waters of our globe.'

Joseph Conrad

The loss of the four masted barque *Pamir* south-west of the Azores in 1957 marked the end of the age of training while trading under sail. *Pamir*'s sister in this enterprise, the similarly rigged *Passat*, was laid up and destined to be a town and museum ship for the German resort of Travemünde.

Passat.

The idea of a town or indeed a museum ship for that matter, is relatively new. The latter role has, however, provided a proper resting place for such famous vessels as *H.M.S. Victory* at Portsmouth, the Royal Swedish Navy's *Vasa* at Stockholm and *U.S.S. Constitution* at Boston. One of the original adventure training ships, *Joseph Conrad* (ex *Georg Stage I*), still lies alongside at Mystic Seaport, Connecticut. There comes a time when the value of ships as part of the historical record is so important that to use them for training, even in a stationary role, may hazard their eventual survival. In due course a place will have to be found for another such vessel in Portsmouth Harbour, the training ship *T.S. Foudroyant*.

The town ship concept provides a secure berth for ships of note and enables them to continue to play a part in the life of the waterfront by acting as a method of entertaining visitors. *Passat* fulfils both these roles. The full-rigger *A. F. Chapman* (ex *Dunboyne*, *G. D. Kennedy*, launched Whitehaven, England, 1888) is now a youth hostel and usefully decorates the old quays of Stockholm.

The steamship had finally won the day and freight could no longer be relied upon, or indeed hoped for, to provide the necessary subsidy that enabled seamen to be trained while the ship went about her business under sail. The steamers adopted the idea for themselves, teaching their cadets aboard power driven vessels. Some shipowners poured scorn on the value of any form of training under sail and those who saw value in it preferred to leave it to the nautical schools. These colleges and institutions could train boys more economically on the cadet ship, where the schedule was dictated by the needs of training, rather than that of securing and delivering a cargo. Sail training in the form of training while trading had helped the steamship through its most difficult period. The concept had provided skilled men who could stand up to the appalling conditions endured aboard the early power driven vessels. Life aboard those damp, coal dust infected, tramping ships, perhaps up to the First World War, was far more unpleasant than that experienced aboard their sailing contemporaries. Seamanship under sail contributed in no small way to the eventual success of power at sea and, indeed, can continue so to do.

It is possible that the end of classical sail training, training while trading, may be temporary, for with the need to conserve what remains of the world's non-renewable resources, wind power may once again have a place. Ideas such as Professor Prölss' *Dynaship*, a 17,000 ton wind propelled, computer tracked vessel, points this way. There is no reason why these ships should not train while they trade, though such vessels would not be so demanding in either skill or endurance as for example the ships of Laeisz's Flying P Line. There is now a chance, however, that the concept of training while trading, in the old sense, will continue for the four masted barque *Peking*, once the British stationary school ship *Arethusa*, belonging to the Shaftesbury Homes and *Arethusa* Training Ship Scheme, may be converted back to her original role by the South Street Seaport Museum, New York.

The Cadet Ships

Training at sea under sail, therefore, passed from the wind driven cargo vessel to the cadet or school ship. These ships, of course, had been part of the tradition of naval and merchant naval training for a long time.

As Alex Hurst records in his book *Sailing School Ships*: 'When I was serving my time in big merchant, four masters, deep loaded ships which took the solid oceans all over them – we used to sneer at the school ships. "They", we said, "never know the hard weather of the high latitudes, nor know what it is to have wet decks, and their crews seldom have to take in sails for stress of weather. They cruise in the summer time and have so many boys that they never have the test of working aloft on their own or with few others, while their masting is too light and their sails so small that they are more suited for our pocket handkerchiefs."

'It was only when we sat for our tickets that we saw the fallacy of these opinions. What did we know of some examination subjects from practical experience? For the time we spent in sail, the answer is "very little".'

This sums up the difference between classical sail training and the concept of the cadet ship, whether designed to provide trained seamen for defence or for trade. It also gives a clear view of the advantages of the cadet ship. This type of vessel can be designed to provide practical experience of all the needs of the mariner, as well as ensuring that he has working experience in the environment of his calling.

The Naval and Coast Guard Experience

Modern cadet ships achieve this in various ways. The United States Coast Guard cadets, in their four summers at the New London Academy, usually take two cruises aboard modern, power driven cutters. Their training barque *Eagle* (see page 106), in addition, takes cadets of the first and third class for a two and a half month cruise to Europe, or the Caribbean. On the cadet's first cruise aboard *Eagle*, he performs the duties that enlisted men carry out on the average, all-power Coast Guard cutter. He does spells as helmsman, look out, signaller, messenger, oiler and the multitude of duties that are possible aboard a ship that possesses not only the means to perform well under sail, but the machinery for doing almost the same under power, with the most modern equipment. On his second cruise, the cadet is an upper classman and undertakes the duties of an officer, being in charge of the deck, engine room or communications.

In addition to the long cruises, 'swab' classes take short, introductory cruises along the eastern seaboard and out into the western Atlantic.

It is aboard *Eagle* that cadets prove their ability. They learn by doing, and understand the principle of leadership by being forced to rise to the occasion. The 179 cadets aboard realise that they are the working part of the ship and that she cannot be persuaded to go anywhere before they have literally learnt to locate every one of the 154 leads in the dark.

Eagle shows how the wind, weather and sea affect a vessel and this lesson is learnt in a way that is difficult to forget – it is physically experienced by the cadet himself, on or above the deck. The basic arts of the mariner, such as knots and splices, chipping and painting, are practised, not as an exercise, but as a necessary part of the ship's life. Equally important, the cadets learn to live together under pressure of work that is seen and understood to be essential. This is not the routine of the power vessel, whose working machinery is for the most part protected by thick skins of metal and plastic and whose lifelines are covered electric wires and hydraulic tubes. The cadet ship has her most vulnerable parts out in the elements, the working environment, and the exposed lifelines are those that support and work the sails that propel the ship.

The Coast Guard is not alone among the American defence forces in believing in the value of sail training – the United States Navy at Annapolis has the 100 ton schooner *Freedom* built in 1931, as well as a number of yachts.

The Federal German Navy uses its training barque, *Gorch Fock II*, which is similar in design and size to *Eagle*, in much the same way, undertaking two cruises, one short in the Baltic and European waters and the other trans-Atlantic.

The Royal Netherlands Navy employs a relatively tiny vessel, the ketch *Urania* which started life as the yacht *Tromp*, for midshipmen training, while the Swedes also use small vessels, the schooners *Falken* and *Gladan* (see page 74). The French have perhaps the most beautiful small ships of all – the twin schooners *La Belle Poule* and *L'Etoile*. These five vessels have for many years appeared to be the ideal way of combining the advantages of sail training with the view that seamen of a modern navy need to be trained also in the fighting ships of the period. This argument loses some of its importance however when the results of the training voyages of *Gorch Fock II* and *Eagle*, particularly the economy of operation and their effective 'showing the flag', are considered.

On the whole, the British have maintained the view of Captain Penrose Fitzgerald, R.N., who violently attacked the value of sail training in 1882. It is, however, carried on at the Britannia Royal Naval College and the Royal Naval Engineering College in such 13 ton yachts as *Pegasus* and *Gawaine*, both sloop-rigged. The Joint Services Sailing Centre at Portsmouth now has a number of specially built yachts – the ketch *Sabre* and the cutter *Adventure* are examples. The Royal Danish Naval Academy has two bermudan yawls of 52 tons each, *Svanen* and *Thyra*, which are employed in much the same sort of seamanship training. With naval craft becoming smaller and, indeed, much more effective, there is an argument for better understanding of the sea at closer quarters. This could perhaps be better done in the Swedish, Dutch or French manner than in a much smaller yacht.

Henry the Navigator founded the first Portuguese Navigation School and his bust decorates the bow of its naval training barque *Sagres II* (see page 118). She was built to the same design as *Gorch Fock II* and *Eagle* by the same builders.

The Spanish Navy owns the four masted, topsail schooner *Juan Sebastian de Elcano* (see page 60) whose near sister ship, the barquentine *Esmeralda* (see page 132), serves the Chilean Navy. Both these vessels undertake world-wide cruises, *Esmeralda* vying with the Argentine full-rigged ship *Libertad* (see page 120), for the largest number of nautical miles sailed while training each year.

Further north, the Colombian full-rigged ship *Gloria* (see page 134) trains the Colombian Navy, but the Brazilians have converted their 2,815 ton displacement, four masted, topsail schooner *Almirante Saldanha* into a motor vessel for oceanographic work and no longer have a large sail training vessel.

The largest cadet ship in the Mediterranean is the Italian Navy's full-rigged ship *Amerigo Vespucci* (see page 58). She is the most easily recognised of all the large training vessels, being built on the lines of a 19th-century line of battle ship, with a gilded stern gallery and an ornate bow, finishing in a figurehead of the Florentine navigator. *Amerigo Vespucci* takes her 150 deck and engineering cadets on five month courses in the Mediterranean and the Atlantic.

The Italian Navy's school at La Maddalena, Sardinia,

1. Palinuro.

uses the barquentine *Palinuro* in the Mediterranean. The Italian belief in the effectiveness of sail training is further emphasised by their two modern bermudan, 70 ft (21 m) yawls, *Stella Polare* and *Corsaro II*, both of which take 10 junior officers under training.

The Yugoslav naval officers are trained aboard the three masted, topsail schooner *Jadran*. Her cruises are mainly restricted to the Mediterranean. She carries a large number of cadets, 132 in all, and they are accommodated in hammocks. This is not unusual in cadet ships, for apart from space saving, it is considered a form of training in itself.

Amerigo Vespucci's sister ship, the full-rigged *Dunay*, now belongs to the Russians and is part of a fleet of sailing ships used in their navy and merchant navy training programmes. She is based at Odessa. The four masted barque *Sedov* (ex *Kommodore Johnsen, Magdalene Vinnen*) acts in a similar capacity, training cadets for the navy. The Russians demonstrate clearly their belief in the value of this form of basic instruction, and they have probably expanded their navy to a greater degree than any other country since the end of the last war. A large number of the officers for this expansion have passed through their sail training programme aboard the naval cadet ships. The Poles followed this example and cadets from the Polish Naval Academy are trained aboard the three masted schooner *Iskra*.

However, an understanding of the effectiveness of such courses is not confined to Europe, the Baltic and the Mediterranean, and the barquentine *Dewarutji* of the Indonesian Navy demonstrates this.

Merchant Navy Training Under Sail

With the virtual end of training while trading, the merchant navy turned to the cadet ship. Captain G. W. Wakeford, founder and past principal of the Southampton School of Navigation (see ketch *Moyana*, page 36) used to illustrate the difference between the navy and merchant navy by saying 'that the former were always practising for something which hopefully, they were never called upon to do, while the latter were continually doing what they never had time to practise'.

The navigation schools are as important to the merchant navy as the naval colleges are to defence. Instruction under sail usually takes place during this period. For example, Southampton School of Navigation had the ketch *Moyana* (see page 36), for some time the only sail training ship attached to a British navigation school, and plans for her replacement ship formed the basic design for the Sail Training Association's schooners *Sir Winston Churchill* and *Malcolm Miller* when the school's finances would not allow new building. Captain Christopher Phelan, the present head of the school, believes strongly in the benefits of sail training and serves on the Sail Training Association's committees. His views are explored more fully in the next section. Southampton now relies on the ketch *Halcyon* for training and instruction under sail.

Russia leads the field, however, in the use of sail training with her merchant navy and fishery training programmes. These follow the example of their Navy. The magnificent fishery training barque *Krusenstern* (see page 98), once Ferdinand Laeisz's four master *Padua*, is the largest merchant navy cadet ship in commission. She is longer over-all than the world's biggest, the Argentine Navy's full-rigged ship *Libertad* (see page 120). In addition, the Ministry of Shipping uses *Tovarishch* (ex *Gorch Fock I*, see page 96). The Russian merchant and fisheries services

2. Halcyon.

also have the advantage of a number of smaller vessels, for after the last war, as part of a programme of reparations, the Ministry of Fisheries and the Ministries of Shipping and Mercantile Marine ordered a number of wooden barquentines and schooners from Finland. The barquentines *Alpha, Horisont, Meridian, Sekstant* and the schooner *Kapella* were part of this programme, as were the fishery and oceanographic vessels, the barquentine *Kodor* and the schooner *Zarja*.

3. Meridian alongside, with two sister ships.

It is reported that the Russians are considering ordering new sail training ships of the size of *Gorch Fock II* and this potential move emphasises their belief in the value of this form of preparation for a life at sea. Such courses must be particularly valuable to their large fishery interests, where fleets of trawlers served by factory ships virtually cover the world. The demands made on seamanship and the ability of the crews to live together in harmony in the hostile environment exploited by this work makes training under sail an excellent base for development. Judging by the size of the Russian sail training fleet, this appears to be clearly understood.

The Norwegians, too, with their large merchant navy, have in the past maintained three fine cadet ships. These were the barque *Statsraad Lehmkuhl* (see page 64), the full-rigger *Sorlandet* (see page 70) and the similarly rigged *Christian Radich* (see page 66). Of the three, only the *Christian Radich*, operated by Ostlandets Skoleskib of Oslo, continues as an active training vessel. The problems are, as usual, those of finance. A government subsidy is necessary and this has been concentrated for the time being on the *Christian Radich*, ensuring her future. The other two ships are now privately owned, having been disposed of because of the difficulties in meeting the running costs. The Norwegian belief in the value of sail training was emphasised by King Olav at the end of the 1968 Sail Training Association Tall Ships Race from Gothenburg to Kristiansand.

'Some people will perhaps ask: "Is it really necessary – in these days – to keep up the tradition of training in sailing ships?" I believe the answer is an unconditional yes. It is a fact that it is of immense importance for the youth of today to learn the art of sailing. Why? Because one who has just been through life on a sailing ship or in a sailing yacht will come into contact directly with the forces of nature itself with no assistance from modern mechanics. In the complex, hard and energetic times we are now living in, we human beings have a fundamental need towards getting into contact with the primeval, the simple, the natural and the genuine in our existence.'

Sweden used to use the four masted bermudan-rigged motor schooner *Albatross* for merchant navy navigation and engineering courses as well as for oceanographic research. Now, however, with the exception of the Stockholm Seamen's School's ketch *Lys*, their sea-going sail training vessels are restricted to the Navy's schooners *Falken* and *Gladan* and the adventure training vessels that will be briefly described later. The stationary school ship, the four masted barque *Viking*, is now moored at Gothenburg.

Finland might have been expected to present a different picture, for the Finns did so much, through Erikson of Åland, to keep the flag of classical sail training flying. He owned the famous ex school ships, the four masted barques *L'Avenir* (see page 20), *Viking*, as mentioned above, and *Herzogin Cecilie* (see page 24). Finland retains a full-rigger, *Suomen Joutsen* or *Swan of Finland* (ex *Laennec*, *Oldenburg*) at Turku, as a stationary school ship for the Finnish merchant navy.

1. Suomen Joutsen.

One of the most interesting merchant navy training vessels is the full-rigged ship *Danmark* (see page 52) for she not only gives instruction to her 80 cadets, preparing them to be officers in the Danish merchant navy, but is used to represent Denmark on occasions of national or international importance. This is normally the province of the navy in maritime countries, but the Danish government has come to realise how much more effective its merchant navy training vessel is for these occasions. A flotilla of minesweepers hardly inspires a glance in a modern harbour, while *Danmark*'s 135 ft (41 m) mainmast can be seen streets away and demands attention even from 'an individual whose soul is indissolubly wedded to the pedestrian shore', to paraphrase Conrad. The Danes, in fact, have two other sailing vessels devoted to merchant navy training: the topsail schooner *Lila Dan*, built in 1951 and owned by J. Lauritzen, and the small, full-rigged ship *Georg Stage II*, built by the Frederikshaven Shipyard in 1935, to replace the earlier full-rigger of the same name, purchased by Alan Villiers, and now, as the *Joseph Conrad* (see page 16), a stationary school ship at Mystic Seaport, Connecticut.

2. Georg Stage II.

The 298 ton gross steel *Georg Stage II* is 123 ft (37 m) long with a beam of nearly 30 ft (9 m). She carries 60 cadets and 10 officers and petty officers. The boys, who range from 15 to 18 years of age, sleep in hammocks.

According to Danish law, every young man who wishes to go to sea and serve on deck, must receive three months training at a seamen's college or aboard a cadet ship before signing on a merchant ship. The boys join about the 20th April and the ship cruises in the Baltic and North Sea, perhaps visiting Scotland. *Georg Stage II* returns to Danish waters in August and pays off after a short examination on the 1st September. Some of the cadets will then transfer to *Danmark*.

Seven thousand, three hundred boys have passed through Stiftelsen Georg Stage Minde since its foundation 94 years ago. It is remarkable that this training has been carried out in only two *Georg Stages*, a magnificent tribute to the economy of sail training ships in general and to these two small full-riggers in particular.

1. Seute Deern.

2. Amphitrite.

The Germans owe much of their interest in sail training to the Duke of Oldenburg and his German School Ship Association at the close of the last century. Both the German Navy and merchant service have continued to take an abiding interest, although two wars and the subsequent confiscation deprived them of all their former ships with the exception of *Schulschiff Deutschland*, now a stationary school ship at Bremen. The East Germans have the splendid 290 ton barquentine *Wilhelm Pieck*, built at Warnemunde in the German Democratic Republic in 1951.

The Polish State Sea School at Gdynia owns the full-rigged ship *Dar Pomorza* (ex *Princess Eitel Friedrich*), bought with money raised by public subscription in 1929. She is one of the oldest sail training ships still operating, having been built in Germany by Blohm & Voss in 1909. The Poles also have a large number of smaller vessels, some of which come into the adventure training category.

The lead given by the Royal Netherlands Navy with *Urania* has not been followed by the Dutch merchant service. The merchant navy's barque *Pollux* was not designed to go to sea, but rather to act as a stationary school ship for the Amsterdam Merchant Navy School for Seamen.

The Belgians, on the other hand, have the barquentine *Mercator*. She is owned by the Association Maritime Belge, which in the past owned the four masted barque *L'Avenir* (see page 20). The Belgian Marine also has the 160 ton bermudan ketch *Zanobe Gramme*.

The splendid French naval schooners *La Belle Poule* and *L'Etoile* (see page 114) have no counterpart in the French merchant navy and Spain, too, is without a sail training ship for her merchant marine.

The Portuguese have a great tradition of sail and their Grand Banks schooners could be converted into fine sail training ships for the merchant navy. In fact, the three masted, gaff, cod schooner *Hortense*, was considered as a replacement for the British *Prince Louis II*, now the French *Bel Espoir*, until the Dulverton Trust decided to build the three masted, topsail schooner *Captain Scott* (see page 82). However, in spite of this richness of ships suitable for conversion, the Portuguese merchant marine, unlike their Navy, does not at present own a sea going sail training vessel of any size.

Of the Mediterranean countries, only Italy and Greece train their merchant navies under sail. The Italians use the steel barquentine *Giorgio Cini* (ex *Fantome II*, *Belem*). She is based on the island of San Giorgio, Venice and makes short voyages to the Adriatic. The Greeks, of course, have the *Eugene Eugenides* (ex Flying Clipper *Sunbeam II*). This three masted, topsail schooner belongs to the National Mercantile Marine Academy at Piraeus, making cruises of three months duration during the summer and shorter ones during the winter, and acting as a training vessel for a number of Greek merchant navy schools.

The Greek, German and Japanese examples could be usefully followed in both Britain and America. In the United Kingdom there are a number of nautical colleges, some possessing small sail training vessels like the Southampton School of Navigation's *Halcyon*. Various

1. Eugene Eugenides.

attempts have been made to rationalise merchant navy training in Britain by cutting down the number of nautical colleges. Some success has been achieved in this direction and it was hoped that this might lead to a single-training ship for the merchant navy, each college taking advantage of the facilities so provided. However, little progress towards this ideal has been made.

America, too, could perhaps profit from this idea. The State University of New York (S.U.N.Y.) Maritime College is, in fact, the oldest institution in the United States dedicated to the education of merchant marine officers. The college used to have the sloop of war *U.S.S. St. Mary's* and later the barquentine *U.S.S. Newport*, but it has now decided against the advantages of sail training, and educates its cadets aboard the school ship *Empire State* (ex *U.S.N.S. Barrett*) acquired in 1973, a 17,600 ton former merchant vessel.

The Ocean Academy, however, used to have the brigantine *Albatross* of 93 tons, but she was lost in the Gulf of Mexico in the mid sixties after being struck by a 'white squall'. Such disasters have made authorities almost overparticular when examining ships—or design proposals—before granting the necessary certificates.

Adventure Training Vessels

Most countries subscribe to the idea of adventure training, though few have developed its potential at sea as part of the process of education.

As described in *Ships of the Past*, adventure training owed its beginnings to the work of Dr. Kurt Hahn and perhaps to the splendid voyage round the world by Alan Villiers in the full-rigged ship *Joseph Conrad* (see page 16) before the Second World War. Captain Erikson also took those who looked for the experience, as Eric Newby describes graphically in *The Last Grain Race*. The idea can best be summarised as 'training by the sea rather than for the sea'. All the ships mentioned above have been designed and ordered to the requirements of navy and mercantile marine training. The sea, however, is a very challenging environment and it can be exploited for reasons other than defence and trade.

In Britain, the work of Kurt Hahn led indirectly to Lord Amory's London Sailing Project, which initially made use of the ketch *Rona* (ex *Aura*, *Alver*), a 77 ft (23 m) wooden yacht built in 1895.

The *Rona* idea started after the Second World War and has subsequently been expanded, sailing out of Portsmouth Harbour under the leadership of Lt. Commander Walter Scott, O.B.E., D.S.C., R.N.

The Sail Training Association also believed that the sea could be used to develop character. Adolescent boys and girls, who have more often than not grown up in the close environment of home and school, can form twisted pictures of themselves which may influence their development. By taking them out of this confined atmosphere for a very short period of time, a different view can be provided. The successful child can be shown another side of himself, and those who have lost confidence can regain it. These are qualities which both navy and merchant navy have learnt to value through the cadet ship, and they can be unlocked by the adventure training vessel to youth of all backgrounds and ambitions.

After a considerable struggle against the traditional

2. Rona.

view of the value of sail training held by both the British Merchant and the Royal Navy, the adventure training schooners *Sir Winston Churchill* and *Malcolm Miller* (see page 90) were built in 1966 and 1968 respectively. The Sea Cadet Corps soon followed in 1971 with the brig *Royalist* (see page 86), in which the ideals of adventure training were wedded to the possibility of eventually joining the Royal Navy. In the same year the Dulverton Trust built the three masted topsail schooner *Captain Scott* (see page 82), which has been used to develop the ideals of Kurt Hahn to the full, combining time at sea with adventure training ashore in the challenging environment of west and north Scotland. She is Britain's largest sail training ship actively working, rivalling the barquentine *Regina Maris* (see page 78). Kurt Hahn's Outward Bound movement has, in fact, grown rapidly ashore and has become very much part of the British educational scene, promoted by such ideas as the Duke of Edinburgh's Award, which gives young men and women a ladder of achievement. The expanding activities at sea were a natural development, for Britain is a small island and her uplands are becoming damaged by over-use.

The Ocean Youth Club was started by Christopher Ellis, G.C., and Christopher Courtauld in the yachts *Theodora* and *Duet* in 1960, and has developed its own special course, making use of volunteer skippers. Anyone over the age of 15 who can swim, is able to take advantage of the facilities for a week-end, a week or longer, to go across the Channel or the North Sea visiting other European countries. The crews are usually mixed, unlike those of the Sail Training Association which caters for boys and girls separately, or the Dulverton Trust, aboard the *Captain Scott*, which is exclusively male, except for the ship's cook.

Geoffrey Williams, who won the single-handed trans-Atlantic race in 1968, became the Director of the Club and his efforts inspired and expanded the series of 71 ft (22 m) bermudan, foam sandwich ketches based on his Robert Clark trans-Atlantic winner, *Sir Thomas Lipton*.

Master Builder, one of the Ocean Youth Club ketches.

The Club now owns ten of these and estimates that this enables it to provide a berth for one person in 2,600 of the British population aged between 15 and 23. It now plans to increase the fleet by at least one more regionally stationed boat, and then concentrate on improving organisation and methods. There has been great progress in recent years. It is Geoffrey Williams' view that 'as soon as organisations (or people) stop trying to push forward the frontier of their own development then the reverse happens and they regress'. Some of these ideas will be explored in the final section.

The Irish have developed their adventure training programme under sail, through the enthusiastic leadership of Captain G. F. Healy.

In 1964, the Irish Government bought *Asgard*, a 50 ft (15 m) gaff ketch designed by Colin Archer. This small vessel, built at Larvik, Norway, in 1905, had been a wedding present to Erskine Childers, father of the late President of Ireland and author of the great yachting tale, *The Riddle of the Sands*, from his American parents-in-law.

In 1914, *Asgard* received a consignment of 700 guns from a German tug off the Sandettie light vessel. Childers, a strong Liberal, had planned by this means to bring down the British Conservative government. The idea succeeded, but he was shot in a subsequent quarrel between rival Irish Nationalists.

In 1975, after ten years of work under Captain Healy, as an Irish sail training vessel, *Asgard* was laid up at Malahide. While the Irish government decided on the best means of conserving its historic vessel, it bought an 'interim' vessel, the 30 ton bermudan ketch *Creidne* (ex *Rapparee II*, ex *Calcador II*) and fitted her out as a relief. She was renamed *Creidne* after the daughter of an Irish chief from near Dundalk, whose habits displeased her father to a degree that he turned her out. In order to regain favour, she went raiding the Scots until her success and perhaps spoils changed his mind.

Convinced by *Asgard*'s success and the need to replace her, the Irish Department of Defence—which, with the Ministry of Finance, is responsible, through the cadet training organisation *Coiste an Asgard*, and now the recently formed Irish Sail Training Committee for sail training in Ireland—obtained tenders for an 84 ft (25 m) wooden brigantine to the design and construction of the famous Irish yard of John Tyrrell of Arklow. She will be used for adventure training. As planned, the new vessel has accommodation for 20 trainees and a permanent crew of five. She will have single spar wood masts and be built of iroko to Lloyds' 100 A1 class. Her main engine is specified as a 200 hp Kelvin diesel. The new brigantine will be named *Brendan* (see page 153) after St. Brendan of Clonfert, the navigator, who is said to have died in 564. It is said in *Navigatio Brendan* that he and his monks celebrated each Easter by camping on what they thought was an island but what turned out to be a whale.

However, adventure training is not the exclusive province of the organisation, for private yachtsmen make their contribution. The Sail Training Association has helped pioneer the use of yachts as temporary adventure training vessels. The regulations governing the Tall Ships Races allow yachts to take part, provided they meet the safety regulations and that half the total crew is between the ages of 16 and 26, and under training. Training in this context could mean for Royal Navy or merchant navy, or as part of the education for life, which would capture most people who qualified by age. From the very start of the Tall Ships Races in 1956, private owners have been encouraged to lend their yachts. For example, the three masted fore and aft staysail schooner *Creole* (ex *Vira*), lent by Stavros Niarchos, was one of the 20 competitors in the Torbay to Lisbon Race of that year.

The Sail Training Association now awards T.S. (Training Ship) numbers to those who regularly take part, though only those vessels that are engaged in full-time training could be regarded as adventure training vessels by definition.

One of the most enthusiastic British yachtsmen using his vessel in this way today is Brian Stewart, with his bermudan sloop *Zulu*. Apart from being the Chairman of the Sail Training Association's schooner committee that runs the two, three masted topsail schooners, he has used his yacht as a training vessel for many years and has taken part in nearly all the Sail Training Association's Tall Ships Races.

Brian Stewart's counterpart in America is Barclay Warburton III, with his brigantine *Black Pearl*, built by C. Lincoln Vaughan, Wickford, Rhode Island, in 1951. Barclay Warburton founded the American Sail Training Association in 1973 and took part in the Tall Ships Race of 1972 with *Black Pearl*. The replica schooner *America* was a competitor in the 1974 Tall Ships Race in the Baltic and won Class B. Following her success in the race, she went on a cruise which took in Africa. Then her owner, Pres Blake, announced that he would present her, with an alumni fund, to the U.S. Merchant Navy Academy at Kings Point for use in the midshipman sail training programme.

The schooner *Brilliant* acts as the sea-going arm of Mystic's *Joseph Conrad*. She was presented to them by Briggs Cunningham. Ten young Americans are taken to sea on each cruise and this could be the beginning of a scheme on the lines of the British Sail Training schooners. There is no lack of encouragement from the American Sail Training Association.

The steel schooner *Westward*, owned by the Sailing Education Association (S.E.A.) of Boston, operates a thirteen-week oceanographic course under Corwith Cramer. The Oceanic Society of San Francisco and Connecticut brings ship and organisation together, arranging programmes that include limited adventure training under sail, together with general ecological study. An example of this is their work with the Island Resources Foundation of St. Thomas in the Virgin Islands, where they employ the 81 ft (24 m) schooner *Nathaniel Bowditch*, built in 1922. The idea of combining adventure training under sail with the study of the marine environment is being accepted on both sides of the Atlantic and developments will be examined in the final section.

Another facet of adventure training under sail is being explored by the schooner *Pioneer* under the leadership of Peter Stanford. His drug rehabilitation programme has achieved the lowest rate of recidivism of any of New York's drug cure schemes. This development is echoed in Denmark with the schooner *Fulton*, as described later (see page 56).

The ex Dutch steel pilot schooner *Tabor Boy* was mentioned in *Ships of the Past*. She provides berths for 22 boys under the command of Captain George Glaeser, a graduate of Kings Point. The schooner operates a sea ranger programme during the school holidays of her owners, the Tabor Academy of Marion, Massachusetts. The cruises range from Maine to Bermuda. Her British equivalent would be Gordonstoun School's 66 ft (20 m) bermudan ketch *Sea Spirit*, built for them in 1970.

In Canada, the brigantines *St. Lawrence II* and *Pathfinder* are used by the Toronto Brigantine Incorporated, taking boys and girls between the ages of 14 and 18 for cruises of one week on Lake Ontario during the summer.

After the success of the Sail Training Association races and Operation Sail in 1972 at Kiel, the Germans, on 1st July, 1973, founded Clipper, Deutsches Jurgendwerk zur See, which now operates two 200-tonners on the lines of the Sail Training Association scheme (see page 90). Their first ship was the Ring-Anderson ketch *Seute Deern*, owned by the Deutscher Schulshiff Verein of Bremen and she was augmented in 1974 by the *Amphitrite*, which was designed and built by Camper and Nicholson in 1887. Both ships carry 20-22 trainees with 5-6 volunteer afterguard on Baltic cruises. The volunteer masters and mates are drawn from the German Merchant Navy,two exceptions being Hans Engel and Peter Lohmeyer, both of whom are former captains of the *Gorch Fock II* and have done much to further sail training in Germany.

While the Dutch merchant navy does not at present train under sail, its adventure training schooner *Eendracht* is one of the latest additions to the world's sail training fleet. This two masted schooner has been built by the Nationale Vereniging 'Het Zeilend Zeeship' and is organised on much the same lines as the British schooners *Sir Winston Churchill* and *Malcolm Miller*. Indeed, it would be fair to say that the Dutch vessel was inspired by the success of the Sail Training Association's two three masters.

Astral.

The flag of the Dutch Antilles is flown by the ketch *Astral*, a 98 ft (30 m) steel yacht. She took part in the Tall Ships Races of 1970, 1972 and 1974, and is owned by Cornelius Vanderstar. He normally uses her as a yacht, but takes six trainees to supplement his six permanent crew for the Sail Training Association's events. The use of yachts for this purpose is growing and helps enlarge the sail training fleet, particularly during the Sail Training Association races, so enabling more young people to go to sea.

In 1968, the French youth organisation, Les Amis de Jeudi Dimanche of Paris, bought the Outward Bound schooner *Prince Louis II* (ex *Peder Most, Nette S.*). This three masted, gaff schooner, built by J. Ring-Andersen of Svenburg in 1944, was renamed *Bel Espoir II*. The owners are dedicated to taking children to sea and their organisation derives its name from the fact that French children have half of Thursday off, as well as Sunday. The schooner cruises from its home port of L'Abervrac'h, north-west Brittany, to the British Channel coast, to other European countries, into the Baltic and a good deal further. *Bel Espoir II* has a permanent crew of five and can accommodate 24 cadets. Under her new ownership, a large deck-house has been added. As noted later, one of the problems of the vessel was the lack of accommodation, for the schooner had pioneered the Prince Louis scheme under the Dulverton Trust, and this led to the building of a larger ship, the *Captain Scott*.

1. Bel Espoir II.

The French at one time considered building a 600 ton schooner for young offenders, but opposition from sea ports is understood to have provoked reconsideration.

The Poles probably believe in the use of the sea for adventure training more than most countries and they have a small fleet of yachts devoted to this purpose. The Warsaw Yacht Club of the Polish Students Association has the *Konstanty Maciejewicz*. She was built at Gdansk, Poland in the spring of 1971 and is some 40 ft (12 m) overall, carrying a permanent crew of two and eight trainees. The *Konstanty Maciejewicz* is used for sea training, making seven cruises into the Baltic and two further afield to the North Sea and Atlantic each year. The students learn navigation, pilotage, meteorology, and rope work, in a programme laid down by the Polish Yachting Association. It is necessary in Poland to have a licence to take a yacht to sea and, therefore, after each cruise there are examinations for the five grades – the Captain of all the Seas, the Captain of the Baltic Sea, Marine Yacht Helmsman, Inland Water Yacht Helmsman and Sailor. *Konstanty Maciejewicz* is responsible for the training of around 100 cadets a year. In 1972, the small training ship sailed around South America, leaving Gdynia on the 21st October of that year. She anchored in Bahia Crossley, Staten Island, Cape Horn and was probably the first sailing vessel so to do. She is named after Captain Konstanty Maciejewicz, the first master of the Polish merchant navy full-rigger *Dar Pomorza*.

Another of the fleet is *Hetman*, owned by Jacht Klub Morski 'Kotwico'. The yacht was built in 1936 by Rasmussen of Hamburg for the Polish Navy Yacht Club, Gdynia. It is used as a training vessel for the Poles, visiting European and Baltic countries, as well as the Soviet Union.

2. Konstanty Maciejewicz.

3. Hetman.

The Sail Training School of the Swedish Cruising Club owns three vessels, the ketches *Gratitude*, *Leader* and *Gratia*. *Gratitude* was built at Port Leven, England in 1908 as a Brixham trawler and sold to Sweden in 1930. She was originally equipped with a steam engine. The Swedish Cruising Club bought her in 1957 and put her back to her original rig. There is a permanent crew of six. Training aboard consists of courses for boys and girls between 13 and 18 and includes instruction in theoretical and practical navigation and seamanship. During the year, the club takes 20 trainees on five two-week courses and there are week-end courses for adults during the spring and summer.

Leader also started life as a Brixham trawler. She was sold to Sweden in 1907 and continued fishing until 1930 when she too was given a steam engine, but used as a coastal tramp. The Swedish Cruising Club bought her in 1969 and like *Gratitude* maintains her as a training ship, carrying the same number of cadets and permanent crew. She is older than *Gratitude*, having been built in 1892. The ketch *Gratia of Gothenburg* (ex *Blue Shadow*, *Cinderella*) was built in 1900, again in England. She was brought to Sweden by Einar Hansen of Malmö in 1936 as a yacht, and in 1964 her owner presented her to the Swedish Cruising Club. She carries two fewer trainees than *Gratitude* and *Leader*, but works to the same programme. Her rig was altered from a schooner to a gaff ketch in 1974/75.

The Finns entered the 25 ton bermudan ketch *Merisissi* and the staysail schooners *Meriliisa* and *Susaleen* for the 1972 Tall Ships Race. The Finnish Navy was interested in the British Sea Cadet Corps brig *Royalist* when she was in Helsinki, and it looked at one time as if the Finns were considering building something similar. Indeed, the Colin Mudie-designed *Royalist* has attracted a good deal of attention among those countries looking for a small training ship.

The Danish Museum Service has combined the preservation of a disappearing local craft with adventure training. For the three masted schooner *Fulton* provides a number of unusual courses. One of these is specifically designed to help adult alcoholics and has proved remarkably successful. The development of the specialised course, whether it be for social or ecological studies, as pioneered in Scottish waters by the schooner *Captain Scott* (see page 82) and in New Zealand by *Spirit of Adventure*, seems to offer great opportunity for experiment and development.

The classic sail training ship, the trainer and trader, may have temporarily disappeared, but the cadet ship and the new adventure training vessel have a great deal to contribute, not only to those who go to sea as a career, but for everyone who wishes to see himself temporarily against a different backdrop.

The selection of ships that follows illustrates the variety of vessels that are carrying out today's tasks.

Leader, Gratitude and Gratia, before her rig was altered, sailing in company.

DANMARK

The Danes trace their interest in sail training back to the sloop *Den Dynkerker Bojert* which was sent out on a cruise by King Christian IV (1588–1648). Sail training continued in more recent times aboard the two *Georg Stages* and the four masted barque *Viking*, commissioned in 1907 and sold during the First World War.

Viking was succeeded by the ill-fated five masted barque *Kobenhaven*, which was lost without trace with her crew of 60, including 45 cadets, some time after the 14th December, 1928, when she left Buenos Aires bound for Melbourne. Her loss is one of the mysteries of the sea. It was during the aftermath of this disaster that the government assumed responsibility for training seamen. Discussions with shipowners resulted in the parliamentary decision to build the *Danmark*. The 790 GRT (gross registered tons), steel, three masted, full-rigged ship was designed by Aage Larsen and built by the Nakskove Shipyard, Lolland. She was launched on 19th November, 1932 and in 1933 set off on her first seven months' cruise with 120 boys aboard.

At the time of the German occupation of Denmark in 1940, *Danmark* was at Jacksonville in Florida, having visited the New York World's Fair in company with the Norwegian full-rigged ship *Christian Radich* (see page 66). Most of her cadets joined the Allied war effort and seven of these captained their own ships before they were 21. Fourteen were lost on active service at sea.

After Pearl Harbour, the *Danmark* was handed over to the U.S. Coast Guard at New London by her captain, Knud L. Hansen. The Coast Guard put the ship, her captain and remaining permanent crew to good use – 5,000 cadets were trained by her during the war, using the Danish programme under Captain Hansen. Her cadet complement during this period of service was about 100, a reduction of 20 from her pre-war complement.

Danmark was handed back to the Danish government with grateful thanks on 13th November, 1945 and was soon back training cadets for the Danish Merchant Navy. A commemorative plaque reminds those aboard of her service with the U.S. Coast Guard.

The Americans, much impressed by the ship, had been anxious to buy her and so continue the programme. Captain Hansen, however, pointed out that the German barque *Horst Wessel* was available for the asking as a war prize, and the Coast Guard enthusiastically pursued the idea, sending Commander MacGowan, who had been with the *Danmark* for most of her American service, to collect her from Bremerhaven, where she lay rusting in a bombed-out shipyard.

In 1959, *Danmark* underwent a thorough refit and modernisation, reducing the number of cadets from 120 to 80, although boys still sleep in hammocks and have their sea chests. Captain Hansen retired in 1964, having commanded *Danmark* for 27 years, which may be a record of service in the same sailing ship. He had a considerable and lasting influence on those who served under him and remains a legend among modern training ship captains.

The Danish government realises that the value of the *Danmark* is more than her service to the merchant navy, and the Royal Danish Ministry of Foreign Affairs ensures that the ship represents Denmark on goodwill visits on every possible occasion.

Danmark has certainly done more than most ships in persuading her people and those of other nationalities, particularly the Americans, of the value of training seamen under sail. She, too, has been a frequent competitor in the sail training races.

DANMARK

OWNERS: Danish Goverment
PLACE OF BUILDING: Lolland
RIG: Ship
DRAUGHT (feet): 14·67

BEAM (feet): 32·8

DESIGNER: Aage Larsen
YEAR OF BUILDING: 1933
CONSTRUCTION: Steel
TONNAGE: 845 T.M.

BUILDERS: Nakskove Shipyard
BUILT FOR: Danish Government
L.O.A. (feet): 212·3
PRESENT NATIONALITY: Danish

1. Chips with everything: day work in the galley is part of the training.
2. Elementary ropework: the instinctive ability to tie the right knot quickly and securely is learnt more effectively under sail than through any other form of sea training.
3. Lash up and stow: in 1959, Danmark *underwent a thorough refit and modernization, reducing the number of cadets from 120 to 80. Hammocks and sea chests were retained as this allowed a more flexible use of space below, as well as preserving a link with the past.*

FULTON

The three masted schooner *Fulton* of 102 registered tons, owned by the Danish National Museum, is run as a training ship, a floating museum and a treatment centre. This combination makes her an interesting variation on the adventure training vessel.

Fulton was launched from the yard of C. L. Johansen of Marstal, the home of the wooden trading schooner, in 1915. Her original purpose was to carry cargoes of cement, grain, timber and salt fish for the Newfoundland trade.

The schooner was sold to Sweden as a local carrier of timber, grain and cement in 1923 and fitted with auxiliary power in 1925. In 1960 she returned to Denmark and continued to trade. This long chapter in her career ended in February, 1970, sixty-five years after she was built, when she was bought by Fisker & Nielsen A/S who fitted her out for her new training and conservation purpose and donated her to the National Museum on the 1st May, 1970.

By a mixture of enthusiasm, dedication, generosity and a sprinkling of chance, a seaman, a scientist and an industrialist combined together to give *Fulton* a new lease of life.

Mogens Frohn Nielsen, the seaman, who had been a cadet on *Georg Stage II* and later was mate of the full-rigged ship *Danmark*, knew how life on such a vessel could form character and give rare personal insight. From an early age, he had carefully tended an ambition to run his own training ship.

Ole Crumlin-Pedersen, a naval architect and specialist in nautical archaeology, believed strongly in the idea of a living museum. He had become Director of the Institute of Maritime Archaeology of the National Museum of Denmark, as well as being custodian of the Viking Ship Museum at Roskilde, near Copenhagen. He believed that museums should be used to demonstrate the past in a practical way today and thus act as sources of inspiration to those viewing age-old methods in the present. He was keen to reproduce afloat the experience of recreating an Iron Age village with the aid of historians and school children. Life aboard ship should be lived as it used to be, or as near to that as possible and therefore the ship had to be restored to her original layout, profile and rig.

In 1969 Ole Crumlin-Pedersen had investigated the possibility of buying an old schooner or ketch for the National Museum. Vessels of this general type were becoming scarce, as many had been left to rot, had been broken up or bought out of service as yachts or charter vessels. This search coincided with that undertaken by Captain Nielsen, who was looking for his training ship.

It was the generosity of Mr. E. Fisker of Fisker &Nielsen in offering the Museum 200,000 kr. to buy a ship that made both ideas possible. Captain Nielsen and Crumlin-Pedersen joined forces in January, 1970, selected *Fulton* and acquired her the following month. Mr. Fisker donated a further 100,000 kr. for her re-rigging to the original specification and on 31st May, 1970, the 50th anniversary of the firm of Fisker & Nielsen, the restored *Fulton* was officially handed over to the Danish National Museum.

Thus *Fulton* was transformed from a motor trader to her original 1915 profile. This change made it possible for the Museum to see how she behaved at sea, yet at the same time she was able to provide courses devoted to adventure training under sail.

The first cruise started with students from Copenhagen University on 6th June, 1970. At this time the ship was organised by the National Museum, but from 1971 to 1975 the Viking Ship Museum at Roskilde had charge of her under Captain Nielsen. There were difficulties over finance for both the refit and running expenses which were alleviated in the winter of 1972/73 by a donation of 200,000 kr. generously given for new deck works by Mr. Kerrn-Jespersen, founder of the Civil Engineering firm of Jespersen & Søn A/S who had built the Viking Ship Hall at Roskilde.

'Long term pupils' sail on the schooner as part of a cure for personality disorders, and the ship receives grants from the child welfare service and other institutions to cover their costs. Similar sponsoring can be done privately or even by a trainee on his or her own initiative. A careful mix of trainees is now considered the most successful, including those convicted of criminal offences. This form of recruiting is fairly common among all adventure training ships, though there is a great difference in degree from ship to ship. Similar progress is also being made with handicapped children. Frohn Nielsen has extended this experimental mixing by adding a couple of old age pensioners to a school cruise. After this experience, he said that there is no generation gap, only a generation fear which can be overcome.

The organisation of the ship is novel, for the jobs within the watch are appointed by the members themselves – the ship is, in fact, run by the trainees. Even the job of master is delegated to a boy or girl, under the careful eye of Frohn Nielsen. *Fulton*'s trainees, up to 30 and aged from 15 to 22, are accommodated in the hold in hammocks in accordance with her original layout. The cruises last for one week and are mixed.

Fulton is not just an adventure training ship with experimental ideas, for she can show what it was like to sail aboard a trading vessel in 1915, though an overmanned one. This is an important part of her three-fold purpose.

OWNERS: Danish National Museum
DESIGNER: C. L. Johansen
BUILDERS: C. L. Johansen
PLACE OF BUILDING: Marstal
YEAR OF BUILDING: 1915
BUILT FOR: H. A. Hansen
RIG: 3-masted topsail schooner
CONSTRUCTION: Wood
L.O.A. (feet): 92·35
DRAUGHT (feet): 8·19
BEAM (feet): 23·29
TONNAGE: 199 T.M.
PRESENT NATIONALITY: Danish

1. *Harbour stow:* Fulton *carries a mixed crew.*
2. *More on the foresheet.*

AMERIGO VESPUCCI

The Italian Navy received two full-rigged ships, built on the lines of a 19th century frigate, from the shipyard of Castellammare di Stabia (Napoli). Both were from the design board of the naval architect, Colonel Francesco Rotundi. He was no doubt influenced by the utility of this design in that it offered the best chance of fitting the maximum number of cadets in the shortest water line length. The two ships were *Christoforo Columbo*, built in 1928, and *Amerigo Vespucci* in 1930.

The *Christoforo Columbo* was handed over to the Russians at the close of the Second World War and is now the *Dunay* or *Danube*.

Amerigo Vespucci is the Italian Navy's largest sail training vessel and is, of course, named after the Florentine merchant and adventurer, 1451–1512. Her three decks create a high freeboard and this appearance is further exaggerated by her bridge. Her rig, too, following the 19th century frigate tradition, is tall and narrow. The long bowsprit is made up of jibboom and spritsail yard. As a pre-training aid, the Naval Academy at Leghorn has set up within the school two square-rigged masts and bowsprit, complete with the necessary standing and running rigging. In this way, cadets are able to have a fair idea of the necessary sail drills before going to sea. To make things even easier when they do arrive on board, the pin rails are labelled in brass.

The training programme is designed to serve the needs of the cadets of the first class at the Naval Academy and cruises normally last five months. From her home port of La Spezia, the ship operates in the Mediterranean and the Atlantic. She carries 260 permanent crew and 150 cadets under training.

The *Amerigo Vespucci* underwent a complete refit in 1964 and has two Fiat diesels, coupled to generators that supply twin electric motors. These in turn connect to a single screw through a system of gearing.

With her figurehead of Vespucci, her traditional black and white hull and the gilded stern with gallery, this modern frigate of 3,543 tons displacement attracts considerable attention. She has taken part in sail training races but normally, depending on her programme, attends the start or the finish. Her performance under sail limits her racing capabilities, though there is no doubt that she is the most spectacular training ship afloat today.

The Italian Navy possesses three other sail training vessels, the first of which is the barquentine *Palinuro*, once the French *Jean Marc Aline* and *Commandant Louis Richard*, bought by the Italian Navy in 1951 and named after the helmsman of Aeneas' ship. She is confined to the Mediterranean and has a permanent crew of 61 with 70 trainees.

The two yachts are both 70 ft (21 m) yawls carrying a permanent crew of five, and 10 trainees. Both these vessels have been very successful in the sail training races.

OWNERS: Italian Navy
DESIGNER: Lt.-Col. Francesco Rotundi
BUILDERS: Castellamare Di Stabia Shipyard
PLACE OF BUILDING: Naples
YEAR OF BUILDING: 1931
BUILT FOR: Italian Navy
RIG: Ship
CONSTRUCTION: Steel
L.O.A. (feet): 266·42
DRAUGHT (feet): 23·58
BEAM (feet): 50·42
TONNAGE: 2,686 T.M.
PRESENT NATIONALITY: Italian

1. By numbers: cadets going aloft to dress ship. The Italian Navy is able to pre-train ashore, as the Naval School at Leghorn has two square rigged masts and a bowsprit, complete with standing and running rigging.
2. With her early 19th century appearance Amerigo Vespucci *is the most spectacular training ship afloat today.*
3. Making sail: Amerigo Vespucci *operates out of La Spezia on five-month cruises, training cadets of the first class from the Naval Academy.*

Chasing Gorch Fock 11 *and* La Belle Poule *in the Tall Ships Race, 1964.*

STATSRAAD LEHMKUHL

(ex *Grossherzog Friedrich August*)

The 1,701 ton gross barque *Statsraad Lehmkuhl* was built by J. C. Tecklenborg of Geestemunde in 1914 and named *Grossherzog Friedrich August.*

She was built to the order of the German School Ship Association, which had at that time the ship-rigged *Princess Eitel Friedrich* (now the Polish *Dar Pomorza*, see page 136) and the similarly rigged *Grossherzogin Elizabeth* (now the French Navy's accommodation vessel *Duchesse Anne*).

The new barque was designed as a cadet ship. She has a long poop that stretches almost to the mainmast. The main steering position is on the fore part of this, with double wheels, while similar provision is made aft of the chart house for auxiliary purposes. The deck-house, after the foremast, contains the galley and carpenter's shop. The 180 cadets are accommodated in hammocks on the 'tween deck, with the petty officers and crew forward. The officers' accommodation is on the main deck beneath the poop. There are 24 permanent crew members. The First World War, however, ruined any ambitions the German School Ship Association may have had for her, for she was first confined to short voyages and then handed over to the British as part of war reparations.

In 1922 her British owners disposed of her to the Bergen Steamship Company, which presented her at once to the Bergen School Ship Association as a replacement for its aging barque *Alfen*. *Alfen* had been built as a schooner-rigged steam corvette for the Royal Norwegian Navy in 1877. The new owners altered the name of *Grossherzog Friedrich August* to *Statsraad Lehmkuhl*. This change of fortune for the ship represented another lost opportunity for the British to secure a training ship, either for the Royal or merchant navy. Both navies had set their faces against training under sail and were not to move even when encouraged by public enthusiasm. The Royal Navy in 1936, for example, turned down Lady Houston's generous offer to salvage, repair and convert the splendid barque *Herzogin Cecilie* which had unfortunately run aground off Start Point.

From 1923 to 1939 while under the Norwegian flag, the *Statsraad Lehmkuhl* made long training cruises out of Bergen as part of the Bergen School Ship Association's programme. The cruises started in April and the ship returned to Bergen in the autumn.

In 1940, in common with the full-riggers *Sorlandet* and *Christian Radich*, the barque fell into the hands of the Germans and was turned into a depot ship. After 1945, restoration of the three ships was begun – Norwegian enthusiasm for sail training had certainly not been dampened by the war.

From 1946 to 1949, *Statsraad Lehmkuhl* returned to her pre-war duties but costs were against her, as she was larger and more expensive to maintain than her smaller sisters, the *Christian Radich* and *Sorlandet*. However, the Norwegian Fisheries School chartered her as a stationary school ship, and in 1952, when confidence had returned, she once again set sail with 180 cadets aboard, as part of the training programme of the Bergen School Ship Association. On her 1952 cruise, she sailed into the Port of New York and repeated this again in 1964, as part of 'Operation Sail' that followed the Sail Training Association's race from Lisbon to Bermuda, in which she was a competitor.

However, in 1967, the old financial worries reappeared and the government withdrew its support. It was then that the shipowner Hilmar Reksten of Bergen bought her and embarked on an extensive refit. There was an idea that *Statsraad Lehmkuhl* might form the basis of a university of the sea, operating from the west coast of the United States. However, this idea did not materialise.

In 1972, she acted as flag ship for the hundredth anniversary celebrations of the Bergen Yacht Club, being towed into the tiny harbour at Godoysund, south of Bergen. But for the most part, she has remained at her berth in Bergen under the careful eye of her owner. Her future is still uncertain.

Statsraad Lehmkuhl has a 450 hp diesel as auxiliary power. Her white hull is now decorated with the arms of Norway on her stern, and scrollwork relieves her bow. The figurehead of Grossherzog Friedrich August, the Duke of Oldenburg, that she was given during her building has long since disappeared. Harold Underhill, the great authority on sail training and cadet ships, regarded her as the best looking of all the three masted barques.

OWNERS: Hilmar Reksten
DESIGNER: J. C. Tecklenborg
BUILDERS: J. C. Tecklenborg
PLACE OF BUILDING: Geestemunde
YEAR OF BUILDING: 1914
BUILT FOR: German Schoolship Association
RIG: 3-masted barque
CONSTRUCTION: Steel
L.O.A. (feet): 276·3
DRAUGHT (feet): 17·1
BEAM (feet): 41·5
TONNAGE: 2,012 T.M.
PRESENT NATIONALITY: Norwegian

1. Striking eight bells while at anchor in Bergen Harbour.
2. Close hauled and under full sail. She carries 180 cadets for the Bergen Schoolship Association.

CHRISTIAN RADICH

The Christiania School Ship Association owed its name to Oslo's original name of Christiania and the full-rigged ship *Christian Radich* is the last of a line of training vessels that started with the British barque *Lady Grey* in 1877.

Training in those days was undertaken as part of a Baltic coastal cruise, though at times this schedule was extended to cross the North Sea on a visit to Scotland. It was realised that this was far from ideal from the training point of view and the Association was determined to venture further, if a suitable ship could be found.

In 1889, Christian Radich, a well known Norwegian businessman of Danish origin, died, leaving 50,000 kr. for the purpose of building a training ship. However, he made the proviso that the money should not be used until after the death of his wife and she lived until 1916. This delay was not the disaster it might have seemed, for the sum grew to 108,000 kr. in the meantime. The Association, with public and private help, decided on the purchase of the 1,814 ton steel, full-rigged ship *Mersey*, which had been a White Star training vessel. She was to be renamed *Christian Radich*.

It was intended that this vessel would complement the brig *Statsraad Erichsen*, which would be given the task of carrying out a six months' course for seamen, while the new ship would engage in classical sail training – training while actively trading. In doing this the new vessel would provide a full apprenticeship course leading to mate.

However, the idea did not materialise because of the outbreak of the First World War and the need to restrict her activities, particularly with trainees aboard. *Mersey* was sold at a handsome profit in 1917 without having officially taken on the name of her original benefactor, Christian Radich. The long search for a proper response to this generous gift was to take another eighteen years, for it was not until 1935, forty-six years after the original deed of gift, that the Association laid the keel of the present full-rigged ship, *Christian Radich*, 676 tons.

The new vessel was built by Framnes Mek-Verksted of Sandefjord, to the designs of Captain (E.) Christian Blom and was delivered to her owners in 1937, so relieving the old brig, *Statsraad Erichsen*, now nearly 80 years old. The *Christian Radich* was designed to be manned by six officers, eight ratings and 88 cadets between 15 and 18 years of age, and carried no cargo. The vessel was given an auxiliary engine which could be relied on in periods of calm and for manoeuvring. The purpose was, and continues to be, training for the merchant or Royal Norwegian Navy, though the majority of trainees enter the merchant service. The course was for three months on a schedule worked out by the Ministry of Education.

During the invasion of Norway in 1940, *Christian Radich* was captured at Horten Naval Base in Oslo Fjord. She had just returned from visiting the New York World Fair – had the Norwegian ship delayed her homecoming, in the manner of the Danish full-rigger *Danmark*, she could have avoided what was for her a disastrous war. The Germans used her as a submarine depot ship, after failing to persuade the Association to continue her training role for their benefit. In 1943, *Christian Radich* was towed to Germany, and that was where she was found, in a capsized state in Flensburg Harbour. She was derelict, being little more than a shell, all movable and salvable equipment having disappeared.

It is greatly to the credit of the Association that it was not put off by this set-back, and with the help of the Allies righted her and towed her back to her builders for a complete rebuild. In 1947, she was returned to her owners a new ship, after £70,000 had been spent on her.

Christian Radich then returned to her training programme for her owners, renamed the Østlandets Skoleskib in 1949. The training programme remained the same and is still the three months' course. The owners, however, have managed to fit into this schedule eight of the 10 biennial Tall Ships Races and the *Christian Radich* has a remarkable record, having won the first prize in four of them. Her performance in these races awakened the interest of the Norwegian people in almost the same way that the America's Cup periodically does for yachtsmen in America, Australia, Britain and France. She was extensively refitted in 1963 and given a more powerful auxiliary. The old one was useless for keeping to a schedule, as it would not push her windward in adverse conditions.

Christian Radich's birth was slow and her early years were marred by war, but she is now the only really active member of the fleet of three Norwegian sail training vessels, which included *Sorlandet* and *Statsraad Lehmkuhl* (see pages 70 and 64).

CHRISTIAN RADICH

OWNERS: Oslo Schoolship Association
DESIGNER: Capt. (E.) Christian Blom
BUILDERS: Framnes Mek Verksted A/S
PLACE OF BUILDING: Sandefjord
YEAR OF BUILDING: 1937
BUILT FOR: Oslo Schoolship Association
RIG: Ship
CONSTRUCTION: Steel
L.O.A. (feet): 205·9
DRAUGHT (feet): 15·0
BEAM (feet): 33·4
TONNAGE: 773 T.M.
PRESENT NATIONALITY: Norwegian

1. Off Copenhagen, fore-royal furled.
2. Colours: she carries 88 merchant navy cadets aged from between 15 and 18.
3. Weighing anchor: sea shanties help lighten the work and co-ordinate muscle power.

SORLANDET

The 568 ton displacement, full-rigged ship *Sorlandet* reflects the ideals of her Norwegian sponsor, the shipowner A. O. T. Skjelbred, who established a fund for the building and management of a sail training vessel for the officers of the Royal Norwegian Navy.

One of the conditions of the gift was that the ship should be powered by wind alone. *Sorlandet*, or *Southland*, Seilende Skoleskips, was founded and the responsibility for building the ship given to the Kristiansand South shipyard of Høivolds Mek Versted A/S.

The small, full-rigger was launched in 1927 and is similar in appearance to other training ships of her size and rig, having double topsails, a single topgallant and royals. She has a long poop, midship deck-house and forecastle. Below, *Sorlandet* has accommodation for 90 cadets and a permanent crew of 12, including officers.

The main cruise of the year used to occupy five of the summer months and in 1933 *Sorlandet* visited the Chicago World's Fair. At the outbreak of the Second World War, her owners presented her to the Royal Norwegian Navy. She was captured by the Germans at the Horten Naval Base and towed to Kirkenes in northern Norway in 1942. The occupying forces used her as a detention centre for their own troops. While suffering in this role, she sank, but was refloated and repaired. When the German Navy required a U-boat depot ship at Kristiansand South, she was chosen, and to fit her for this, her masts and spars were removed and a large deck-house was built to accommodate the submarine crews.

In 1947 *Sorlandet* resumed her duties. Eleven-week courses were devised and boys of 15 and over enrolled. Accommodation, classrooms and other facilities were provided at *Sorlandet*'s home port of Kristiansand South. The full-rigger resisted auxiliary power until the winter of 1959/60, when a 240 hp diesel was installed.

Sorlandet has been a frequent competitor in the Tall Ships Races, but her schedule kept her at home for the 1972 and 1974 events. She was sold privately to the shipowner Jan Staubo of Oslo in 1974 and her future is uncertain.

SORLANDET

OWNER: Jan Staubo
DESIGNER: Hoivolds Mek Versted A/S
BUILDERS: Hoivolds Mek Verksted A/S
PLACE OF BUILDING: Kristiansand South
YEAR OF BUILDING: 1927
BUILT FOR: Sorlandets Seilende Skoleskibs Institution
RIG: Ship
CONSTRUCTION: Steel
L.O.A. (feet): 185·3
DRAUGHT (feet): 14·0
BEAM (feet): 29·1
TONNAGE: 644 T.M.
PRESENT NATIONALITY: Norwegian

Sorlandet *is a full rigger with a typical training ship profile.*

FALKEN & GLADAN

The Royal Swedish Navy list of ships used for sail training includes the twin gaff schooners *Falken* and *Gladan*.

The Swedish government announced the decision to build the two schooners in a letter dated the 12th January, 1946. *Falken* and *Gladan* were to follow the classically beautiful full-rigger *Jarramas* of 400 tons displacement. This small ship, in common with her near sister, *Najaden*, and other Scandinavian training vessels of that period – she was built in 1900 – had the look of a naval frigate. *Jarramas* is now a museum ship at Karlskrona, while *Najaden* is maintained in a similar role at Halmstad.

Falken and *Gladan* were designed by Captain (E.) Tore Herlin and the government budgeted $300,000 for their building. *Gladan*'s keel was laid on the 28th May, 1946 at the naval dockyard at Stockholm and she was launched on the 14th November of the same year, the Navy taking delivery, after trials, on the 2nd June, 1947. *Falken* followed immediately, taking advantage of *Gladan*'s stocks. She was launched on the 12th June, 1947, and commissioned on the 1st October of the same year.

Both vessels displaced 220 tons and had the appearance of a yacht with a spoon bow and a counter stern. The sail plan adopted was that of a gaff schooner with topmasts as it was felt that boys learn to handle fore and aft rig in a far shorter time than square rig. They, therefore, have the opportunity of other forms of instruction, the whole course being of short duration. The galley is in the foreward house, while the ward room is in a similar structure abaft of the mainmast. Both ships carry 14 permanent crew and 28 cadets, the latter being accommodated in bunks and hammocks in two flats.

Both schooners have been continuously improved from year to year, the non-commissioned officers, for example, being provided with their own cabins and a mess room. An extensive refit was undertaken in the winter of 1969/70, when two 380/220 k. generators were installed for continous load, replacing the old battery and generator system. This enabled a complete modernisation of the heating, ventilating, cooking, water supply and refrigeration equipment, as well as providing power for modern navigational aids. It did mean, however, that the generators had to be constantly in evidence and it is interesting that this was avoided, to a limited extent, on the three British schooners, *Sir Winston Churchill*, *Malcolm Miller* and *Captain Scott*, by the adoption of the older battery and generator method. It is difficult to generate power on a fully equipped sail training vessel, and still achieve the ideal of quietness at sea.

The naval cadets are aboard *Falken* and *Gladan* for one month, the training season stretching through the five months of summer. Though both ships are mainly confined to the Baltic, they invariably take part in the sail training races.

The Rydberg Foundation chartered *Falken* in 1952 for a training cruise to West Africa and the Mediterranean.

The name *Falken* (or *Falcon*) has a long history. In Gustav Vasa's Navy there was '*White Falken*', '*Calmare Falken*' and '*Finnish Falken*'. The name '*Falken*' alone was used for the first time in 1631, when a 20 gun ship of that name was launched at Stockholm. She was wrecked in 1651. The following year a new 42 gun '*Falken*' was built. She was followed by another built by Karl Karlson at Stralsund in 1689. This '*Falken*' had a crew of 120 men, including 20 soldiers and was some 90 ft (27 m) long. The next ship to bear the name was the sail training brig 'Falken' built in 1877, which ended her training career in 1939 and was eventually wrecked off Juan Fernandez' Island in 1966.

Gladan's name has appeared on the stern of a warship once before. This was the boys' training brig *Gladan* built in 1857 and disposed of in 1927.

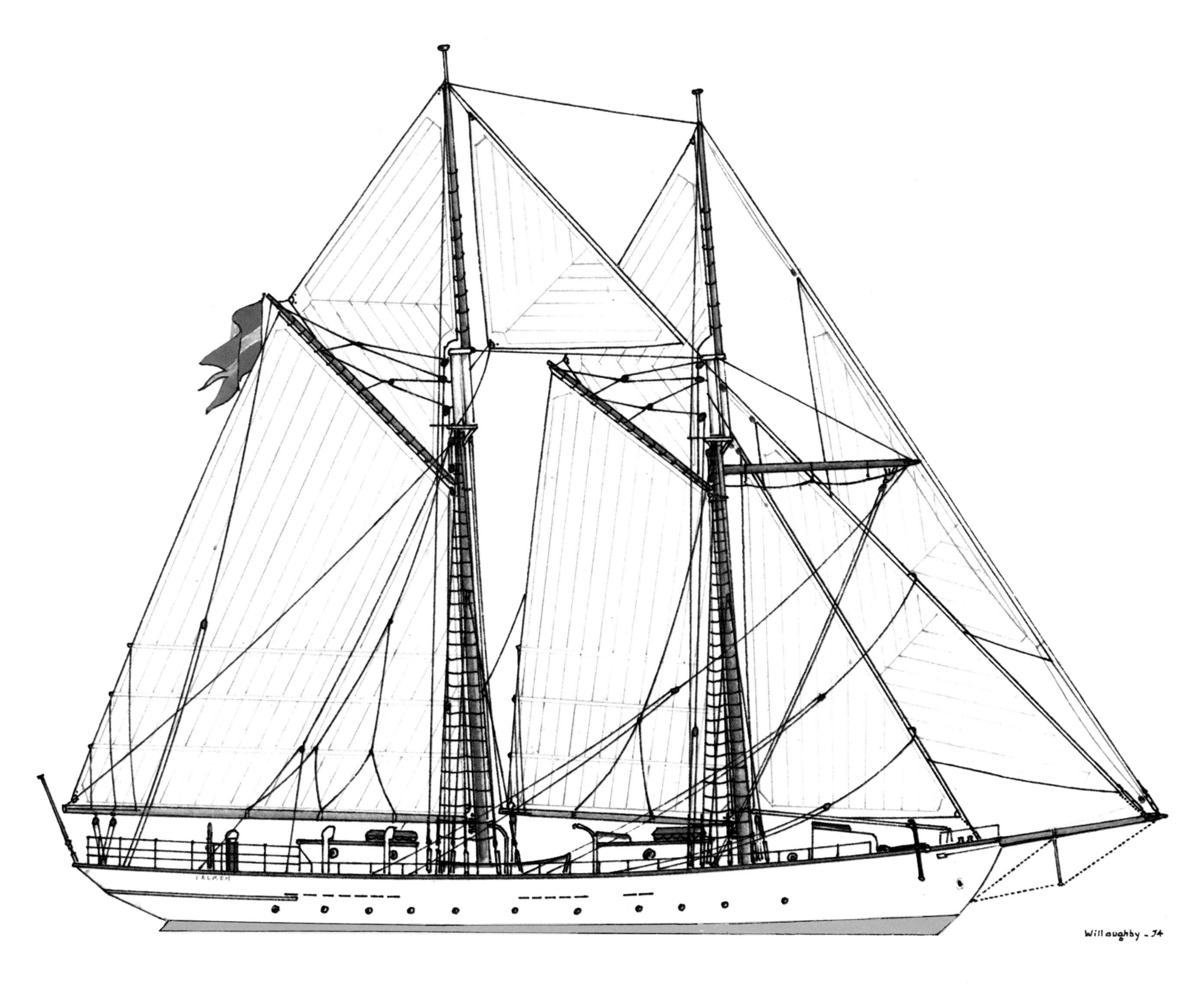

FALKEN

OWNERS: Royal Swedish Navy
PLACE OF BUILDING: Stockholm
RIG: Schooner
DRAUGHT (feet): 13·5

BEAM (feet): 23·83

DESIGNER: Capt. (E.) Tore Herlin
YEAR OF BUILDING: *Gladan* 1946, *Falken* 1947
CONSTRUCTION: Steel
TONNAGE: 232 T.M.

BUILDERS: Naval Dockyard
BUILT FOR: Royal Swedish Navy
L.O.A. (feet): 112·5
PRESENT NATIONALITY: Swedish

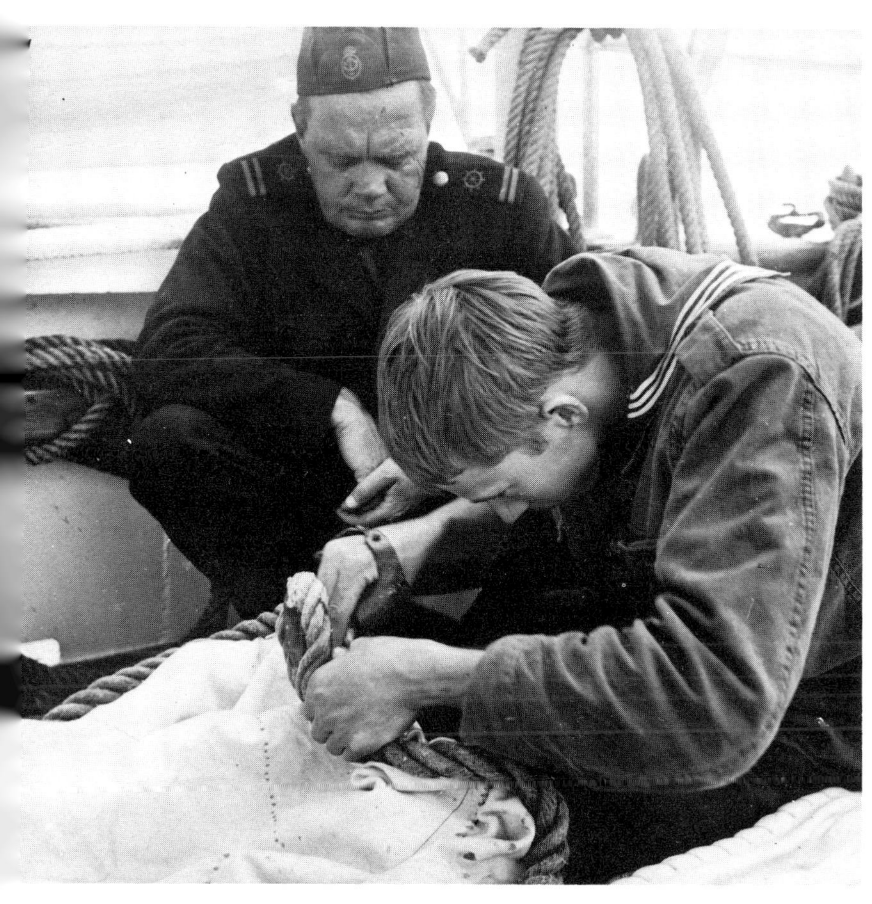

1. *Learning the art of sail-making with needle and palm.*
2. *Cleaning the day's catch: fresh fish caught with a spinner, or by 'ripping' when at anchor, help supplement the normal diet.*
3. *Collision course? Taking a bearing on an approaching vessel with a pelorus.*
4. *Entering harbour under sail with the crew lining the fo'c'slc and the Officer of the Watch saluting.*

REGINA MARIS

The three masted, wooden barquentine, *Regina Maris*, was built as *Regina* in 1908, by J. Ring-Andersen, the famous wooden shipbuilders, of Svendborg, Denmark, to the order of P. Reinhold of Raa in Sweden.

Regina was particularly strongly built of oak planking on oak frames, as her purpose was the Icelandic cod trade, bringing cargoes of dried and salted fish for the Scandinavian and German markets. As was customary, the spaces between the frames were filled with salt to pickle the timbers, and the whole was sealed in by a wooden lining.

After many years at this trade, she transferred to the nitrate business, picking up cargoes from Laeisz's P ships in Hamburg for the small Norwegian, Swedish, Danish and Finnish ports which were closed to the larger vessels. In the early 1930s, she was provided with her first auxiliary engine and her rig cut down to that of a schooner in traditional Baltic fashion.

After the Second World War, *Regina* engaged in general cargo handling, until damaged by fire in 1962. In 1965 she was bought by the Norwegian shipowners Siegfried and John Aage Wilson of the Wilson Line of Arendal and rebuilt at considerable expense by the Kristiansand South yard of Høivolds Mek Verksted A/S.

The work was done to Lloyds' specifications and she is still classed by them. The opportunity was taken to rename her *Regina Maris* (*Queen of the Sea*) and she was rerigged as a barquentine, with double topsails, top gallant and royal with six stunsails, providing 7,200 sq ft (669 sq m) of canvas in all. Because of her change of rig, new sails of flax, with the exception of two jibs and the mainsail which are terylene, were required. A 250 hp G.M. main engine with a 4.5:1 reduction, exhausting through the cap of the lower mizzen mast, was also provided together with new auxiliary services.

Below, the main saloon is aft and is panelled in traditional style. The crew messroom and galley are housed around the foremast on deck. The officers live aft, together with the cook and engineers – the permanent crew of six is accommodated in four and two berth cabins forward. In between there are three four, two three and five two berth cabins off a centre companionway.

The Wilson brothers had bought and rebuilt her to fulfil an ambition of sailing their own vessel round the Horn. They achieved this in 1968 and in fact circumnavigated the world. In Chile, they discovered one of the anchors of Darwin's *Beagle* and brought it back to Norway where it is displayed at John Aage Wilson's home.

Regina Maris became British in 1973 and was sailed from Ensenada, on the west coast of Mexico, to Plymouth by Captain Mike Willoughby. On this voyage of 12,000 miles the barquentine averaged 5 knots. She arrived back in time to compete in the 1974 Tall Ships Race from Corunna in north Spain to Portsmouth, England.

In the summer of 1975 *Regina Maris* was chartered by a Polish film company and a British television company to appear in a feature film based on Conrad's 'Shadow Line'. For this purpose she was refitted at Camper and Nicholson and then sailed, again under the command of Captain Mike Willoughby, to Baljik on the Black Sea where the filming took place.

Subsequently, she was sold to the Ocean Research and Education Society founded by the publisher of 'Sail' magazine, Bernard Goldhirsh, and to Dr. George Nichols, Jnr., a founder of the Sea Education Association (see page 112). Under their ownership she will engage in a study of the distribution, behaviour and food preferences of whales and porpoises.

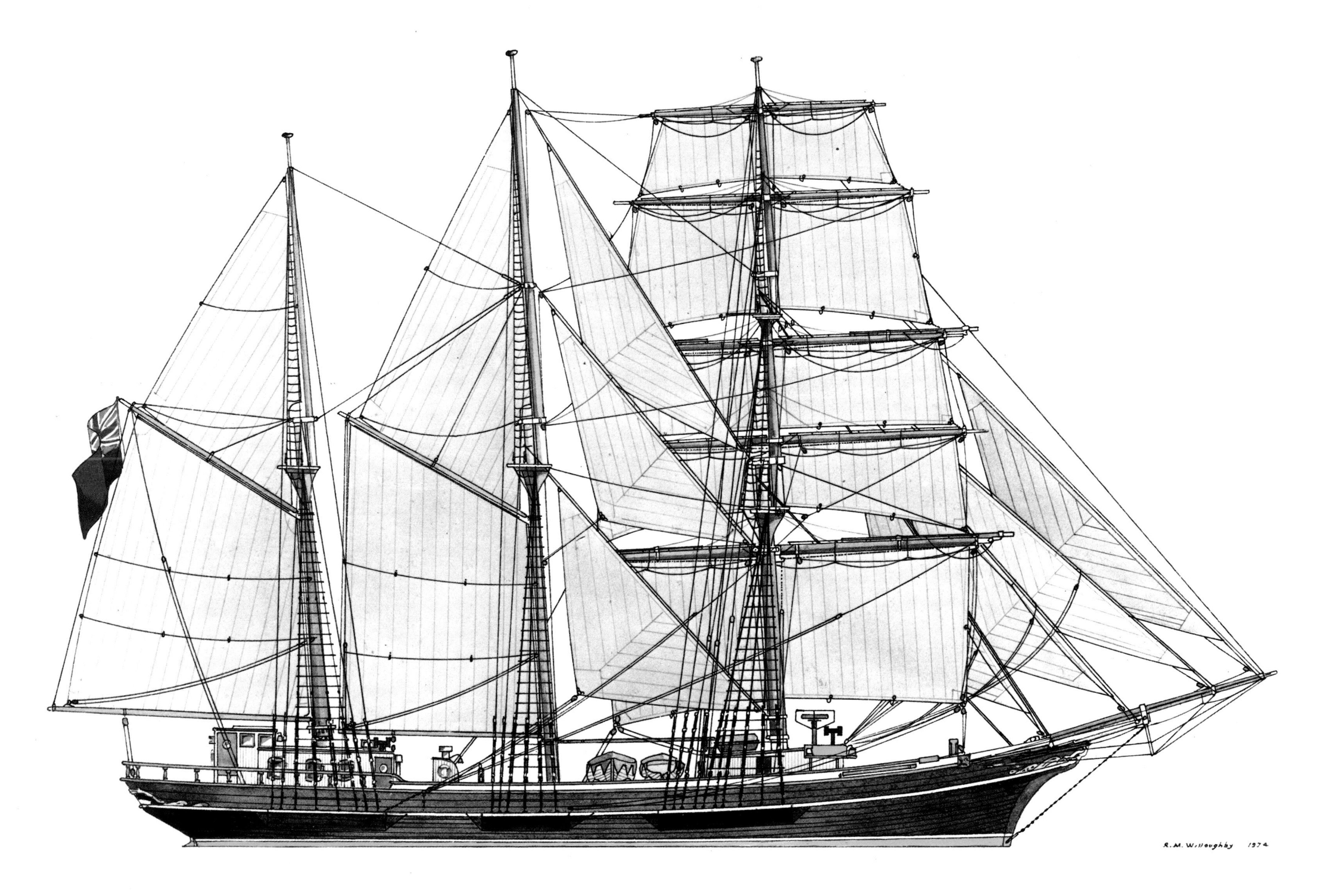

REGINA MARIS

OWNERS: Brictec Finance
PLACE OF BUILDING: Svendborg
RIG: 3-masted barquentine
DRAUGHT (feet): 9·75

BEAM (feet): 25·5

DESIGNER: J. Ring Anderson
YEAR OF BUILDING: 1908
CONSTRUCTION: Wood
TONNAGE: 297 T.M.

BUILDERS: J. Ring Anderson
BUILT FOR: P. Rheinhold
L.O.A. (feet): 118·6
PRESENT NATIONALITY: British

1. A girl trainee under the watchful eye of the bo'sun (centre) and the third mate during the 1974 Tall Ships Race.
2. A mixed crew at work: she sailed 12,000 nautical miles, averaging 5 knots, from Ensenada, West Coast of Mexico to Plymouth, England, in order to take part in the 1974 Sail Training Association Tall Ship's Race.
3. Refit after rebuilding in Kristiansand during the winter of 1965/66: rigged originally as a barquentine, she eventually became a motor schooner, but was re-rigged again by Captain Siegfried and Captain John Wilson as a barquentine, to sail round the Horn.
4. Regina Maris *under sail: she was ideally suited for the leading part in the television film of Joseph Conrad's 'Shadow Line' which was shot late in 1975 in the Black Sea. She was under the command of Captain Mike Willoughby.*

CAPTAIN SCOTT

In 1940, at the request of Dr. Kurt Hahn, the educational philosopher, Captain G. W. Wakeford, O.B.E., F.R.I.N., organised and ran the first course of what later became the Outward Bound Sea School. From this developed in 1948 the Outward Bound Moray Sea School at Burghead, in Scotland.

The connection between the thought behind Gordonstoun School, of which Dr. Hahn was headmaster, and the Outward Bound movement was strong, and it was continued in their ships. The ketch *Prince Louis I* (ex *Maisie Graham*, see page 13) which came from Gordonstoun to the Moray Sea School, initiated the courses that were developed through *Prince Louis II* (ex Danish *Peder Most*, now the French *Bel Espoir*, see page 114), to those now most successfully run by the new, three masted, topsail schooner *Captain Scott*.

Kurt Hahn believed that there was a third dimension to education, adding to the first and second – that is to say, the three R's, reading, writing and 'rithmetic and the development of the body – a third, and perhaps the most important – the encouragement of personality. He believed that adventure and adversity, experienced singly or more particularly in a group, could develop character in a way that would prove lasting and useful.

His ideas were put into practice and developed at sea by Commander Victor Clark, D.S.C., R.N., the last master of the *Prince Louis II*, who had sailed round the world in his yacht *Solace*, soon after the Second World War.

In 1966, Outward Bound decided to dispose of the *Prince Louis II* for economic reasons. However, the movement was persuaded by Dr. Hahn, Commander Clark and others to charter her for six months to the Dulverton Trust as an experiment, to see whether the ideas developed by Commander Clark from the original Outward Bound scheme, would work. Lt. Commander Patrick Job took a prominent part in the original planning and subsequent administration at Plockton.

The scheme was a success in all but its financial aspects, and the Sail Training Association was called in to advise. They discovered that the *Prince Louis'* 24 cadets were not enough to make the scheme viable and suggested a larger ship.

Captain Scott was designed to stay at sea in any weather, summer or winter, that the north and west of Scotland could provide, to be sailed by boys with no previous experience around that rugged coast, to points from which the mountain and hill expeditions would be launched. The schooner had to carry all the necessary gear for these shore parties and to be operationally independent of any fixed base other than the administrative office. The change from sea to land and vice versa was considered an excellent way of developing the Hahn philosophy.

The initial designs were undertaken by the British designer Robert Clark, who had created the three masted bermudan schooner *Carita*. The final rig was developed by the illustrator of this book, Captain Mike Willoughby, late master of the *Sir Winston Churchill* and an adviser to Britain's Maritime Trust.

The standing topgallant yard rig (fore course, lower and upper topsails and topgallant), with fore staysail, three jibs and main topmast staysail, was adopted to provide the maximum work aloft and outboard practicable considering the duration of the course and the cruising area. Gaff sails on all three masts with three rows of reef points and gaff topsails on main and mizzen masts ensure full employment, plenty of hard work and safety in all weathers.

The famous Scottish fishing boat builders, Herd & Mackenzie of Buckie were selected to build this wooden vessel and she was launched by Lady Jane Nelson on the 7th September, 1971. On October the 18th, *Captain Scott*, with a full complement of boys, set off for her home port of Plockton on the Scottish mainland.

The permanent crew consists of her master, mate and deputy master, chief expeditions officer, boatswain, chief engineer, second expeditions officer and cook/caterer, now traditionally a girl. Three other temporary officers are carried on each cruise to act as watch instructors. Thirty-six trainees, young men aged between 16 and 21, are drawn from Britain and overseas. They are not aboard to learn the skills of the mariner nor for a career at sea. Their month's training has relevance to all walks of life.

The trainees are accommodated in bunks on the half deck. The permanent crew live abaft this in separate cabins, leading off their ward room. The master has his own quarters aft immediately below the chart room and wheel-house. It is from this wheel-house that the twin, 230 hp Gardner diesel engines are operated with bridge controls.

The nine cruises each year last for 26 days, beginning at the end of January and carrying on until just before Christmas. Each cruise has built into it three land expeditions, which are spaced through the month, each designed to train for the next. These expeditions are classified as 'Gold' standard for training for the Duke of Edinburgh's Award Scheme, and the whole course counts as residential qualification for the Gold Award. The cruising and expedition area encircles the north-western Celtic fringe, stretching from Northern Ireland in the south, around St. Kilda in the Outer Hebrides to the Orkneys, Fair Isle and the Shetlands in the north. These waters are rich in both a variety of weather and bird life and the ship has undertaken at least one cruise to explore them, with the idea of promoting a better understanding of the marine environment. This is to be expected of a vessel owned by an organisation whose president is the naturalist, Sir Peter Scott. The schooner bears the name of his father, Captain Robert Falcon Scott, R.N. (1868–1912). The words from Tennyson's poem *Ulysses* which are carved on the cross erected in his memory in the Antarctic overlooking Winter Quarters Bay, 'To strive, to seek, to find, and not to yield' decorate the wheel-house, epitomizing the object of the course.

The figurehead shows Captain Scott as he is remembesed, in snow goggles, armed with a ski stick on his famous but fatal expedition to the South Pole. He died in 1912 while leading the British Antartic expedition of 1910–1913 in an effort to be the first men to reach the pole. He was forestalled by the Norwegian Amundsen.

CAPTAIN SCOTT

OWNERS: Dulverton Trust
PLACE OF BUILDING: Buckie
RIG: 3-masted topsail schooner
DRAUGHT (feet): 13·5

DESIGNER: Robert Clark, Commander Clark and Capt. Willoughby
YEAR OF BUILDING: 1971
CONSTRUCTION: Wood
BEAM (feet): 28·0
TONNAGE: 380 T.M.

BUILDERS: Herd & MacKenzie
BUILT FOR: The Dulverton Trust
L.O.A. (feet): 144·25
PRESENT NATIONALITY: British

1. Stowing the main: her crew of 36 boys, aged from 16 to 21, spend 26 days aboard, cruising off Scotland's challenging west coast and taking part in shore expeditions.
2. In a light breeze among the Hebrides.
3. Taking an anchor bearing.

ROYALIST

Encouraged by the success of the 75 ft (23 m) chartered brigantine *Centurion* (ex *Aegean, Beegie*), which won her class in the 1966 Sail Training Association's Falmouth to the Skaw race, the Sea Cadet Corps began planning their own offshore training vessel. The Corps is nationally organised, having units throughout the United Kingdom. Training is normally carried out in small craft, great use being made of ex Navy M.F.V.'s, sailing boats and their two Admiralty 16 ft (5 m) Sea Cadet dinghies.

As part of the design studies, the Sea Cadet Corps chartered other yachts. They were looking for a vessel 'that would train officers and cadets under conditions which would give to the individuals concentrated training in the technical aspect of sailing, engender an 'esprit de corps', and teach the benefits of a disciplined living at sea'. It also had to accommodate the maximum number of Sea Cadet officers and Sea and Marine Cadets. Occasional all-girl cruises were also planned for the Girls Nautical Training Corps, to take them offshore.

To achieve this, they sought the advice of the naval architect Colin Mudie, probably the most imaginative of all today's British yacht designers, and asked him to produce drawings for a ship based on their experience and experiments. It is interesting to discover that he finally produced a small, modern equivalent of the brig that Admiral Lord Cunningham had said 'produced lads unsurpassed for smartness, alertness and physique'. He was referring to his own experiences on the brig of *H.M.S. Martin* in which he had served in 1901.

The new vessel was launched on the 12th July, 1971, and named *Royalist* at a ceremony held at Cowes on the 3rd August by Her Royal Highness The Princess Anne. Her father, Prince Philip, had been the first cadet aboard the adventure training vessel *Prince Louis I* and is now the Admiral of the Sea Cadet Corps. The name *Royalist* has been used by the Royal Navy since 1797.

Royalist was built by Groves & Guttridge of Cowes, Isle of Wight, and was awarded the Lloyds Register Yacht Award for 1971, being considered 'the best constructed, best designed and best equipped vessel for its purpose'. The brig is a combination of the cadet and adventure training ship, for not all her boys are destined for the Royal Navy, merchant service or the fishing fleet. She is the first brig to be built for the British flag for half a century, her rig being chosen for almost the same reasons as those that had decided the Royal Navy on the design at the end of the last century. Square-rig gives *Royalist*'s complement of 32 (24 cadets) the best possible practical experience, necessitating maximum employment and so creating self-reliance. The small size of the sails, set on pole masts, however, ensures that they are within the capacity of the crew, whose ages range from 13 to 18 years.

Below, the cadets are accommodated in two compartments. The mess deck provides for 22, while forward of this there is room for four other crew members. Each trainee has his own bunk and locker. A few of the Girls Nautical Training Corps are also included in some crews and on these occasions they are berthed in the wardroom, though this, with the two two-berth cabins off it, is usually the domain of the local Sea Cadet officers whose boys are aboard.

Royalist carries a permanent crew of five – captain, sailing master, engineer, boatswain and coxswain, the first three of whom have their own cabins. The cook comes on a temporary basis and has a cabin close by the engine room.

Like the two Sail Training Association's schooners, *Royalist* has twin screws. She is powered by two 115 hp Perkins diesels, which give her 6 knots with a maximum in still water of 8, if required, for short periods. Electrical power for her lighting and full navigational equipment is provided by two alternators coupled to two Petter diesels through batteries.

From time to time the brig sails round the British Isles to be as near as possible to the units which she serves. The cruises last a week with the exception of two of 10–14 days during the summer months, when the ship is likely to visit the Continent. Week-end familiarisation courses are also arranged for adults, allowing Sea Cadet officers the opportunity of understanding the vessel before bringing their trainees aboard. Seven hundred trainees pass through the ship in a year, and *Royalist* covers some 6,000 to 8,000 miles.

Under sail, though not close winded, *Royalist* has proved herself flexible, with her large variety of sails, and easy to handle. She was designed to achieve 12 knots under sail and has exceeded this speed on a number of occasions.

Wherever she goes, she attracts a good deal of attention. Painted in the manner of a Blackwall frigate, she looks larger than her dimensions suggest. The main interest, however, is the combination of square rig and a small, efficient hull.

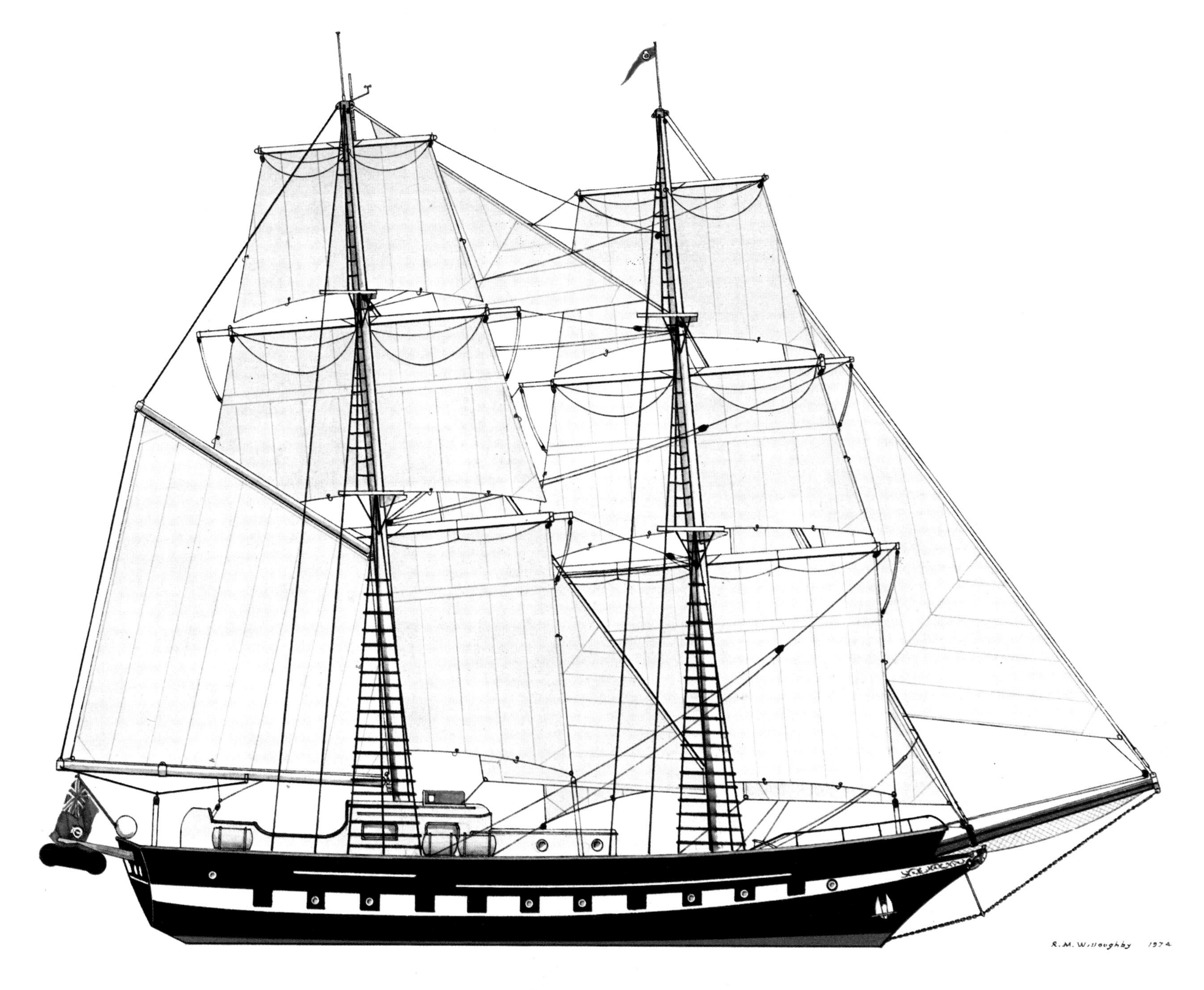

ROYALIST

OWNERS: Sea Cadet Corps
DESIGNER: Colin Mudie
BUILDERS: Groves & Guttridge
PLACE OF BUILDING: Cowes
YEAR OF BUILDING: 1971
BUILT FOR: The Sea Cadet Corps
RIG: Brig
CONSTRUCTION: Steel
L.O.A. (feet): 76·5
DRAUGHT (feet): 9·0
BEAM (feet): 19·5
TONNAGE: 110 T.M.
PRESENT NATIONALITY: British

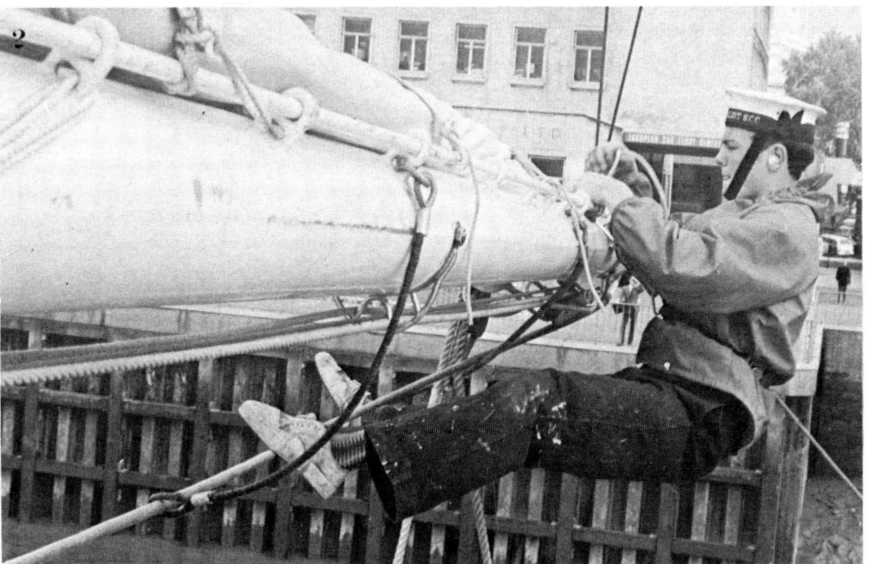

1. The pinrail.
2. Securing the lacings on the spanker boom.
3. Preparing the square sails for setting.
4. Passing a message on the radio.

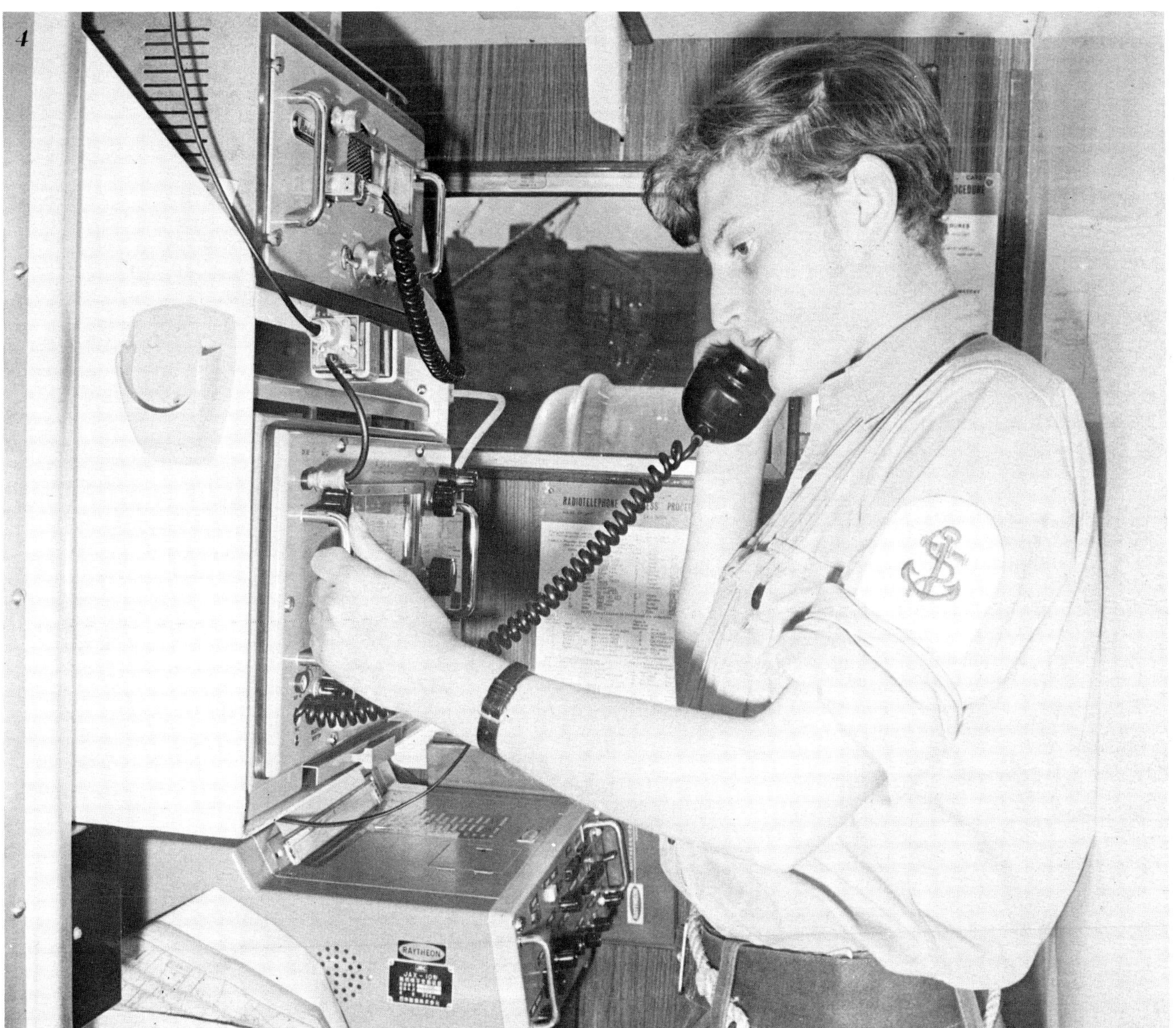

SIR WINSTON CHURCHILL & MALCOLM MILLER

Britain's two 235 ton displacement, three masted, topsail schooners *Sir Winston Churchill* and *Malcolm Miller*, were built for the Sail Training Association in 1966 and 1968 respectively. The Association was the idea of a London solicitor, Bernard Morgan, and it was founded in 1955 to promote the International Tall Ships Race for the world's sail training ships. The first race from Torbay to Lisbon was organised in 1956. Events have been held every two years since that date.

The experience of organising and running the Tall Ships Race soon convinced the Association of the worth of sail training. In 1960 they decided to explore the possibilities of building a British sail training ship. The Royal Navy and British merchant service had turned their backs on this form of training and with the notable exceptions of the Southampton School of Navigation's *Moyana* (see page 36), the Moray Sea School's *Prince Louis I* (see page 131) and Lord Amory's Rona Trust, operating the 77 ft (23 m) ketch *Rona* (1895, see page 47), there was little evidence of British interest in sail training. Unlike Russia and America, for example, which both still made use of training under sail, Britain's Navy and commercial interests had forgotten the benefits well known in her past. The purpose of the new ship was to develop Kurt Hahn's idea of benefiting the whole individual rather than just developing maritime skills.

After turning down the idea of converting an existing vessel, the design of a ship was the first step. The sizes ranged from the 70 ft (21 m) yacht, such as that which successfully started the Rona scheme and was carried on with great success by the Ocean Youth Club (see page 48) to large ships such as the German Navy's barque *Gorch Fock II* (see page 126). However, the duration of the cruise, fixed at a fortnight, the need to attract first class permanent crew and the requirement to work the year round, decided the Association on a 235 ton displacement, 135 ft (41 m), three masted schooner with course, square topsail and raffee on the fore.

The *Sir Winston Churchill* that resulted from these arguments and calculations was designed by Camper & Nicholson in association with Captain John Illingworth, F.I.N., R.N. Camper & Nicholson had been working on the drawings for a replacement of the Southampton School of Navigation's *Moyana* (see page 36). They based their designs for both ships on *Sonia II*, a 166 ft (51 m), three masted schooner, built in 1931 as a yacht for Miss Marion Betty Carstairs, a devotee of both motor boat and car racing. Camper & Nicholson were not strangers to the training ship world for they had designed the Spanish *Juan Sebastian de Elcano* (1927, see page 60), the Chilean *Esmeralda* (1952, see page 132) and the schooner, later barquentine, *Amphitrite* (1887, see pages 46 and 49).

The *Sir Winston Churchill* was built by the Hessle, Hull shipbuilders, Richard Dunston and launched into the Humber on the 5th February, 1966. This event had been temporarily delayed due to a freak storm that had managed to trigger the slipway mechanism and blow her over just before the launch, while she was still on the stocks. Damage to the three masts was considerable, but there were only minor problems with the hull.

The *Sir Winston Churchill*'s 36 trainees, later raised to 39, are organised into three watches, and accommodated in berths on the half deck that stretches from the paint locker and heads forward, to the engine room bulkhead amidships. The engine room houses two 135 hp Perkins diesel engines, driving controllable pitch propellers and two generators associated with a bank of batteries to reduce engine disturbance while under sail.

The after guard is made up of the master, the first to hold the post being Captain Glyn Griffiths, chief officer, engineer, boatswain and cook. The remaining six are volunteers. They are a navigator, who has to be qualified, three watch officers and two pursers.

The master, first officer and volunteers live aft of the engine room, using the chart room in the aft end of the deck-house as their ward room. The chief engineer, boatswain and cook have a flat aft, just forward of the volunteer watch officers' cabin.

The *Sir Winston Churchill* captured the imagination, and confounded any critics that remained. Her success was soon followed by her near sister ship, *Malcolm Miller*, made possible by the generosity of the then Lord Mayor of London and former Lord Provost of Edinburgh, Sir James Miller, whose son, after whom the schooner was named, died in a car accident. Sir James presented half the money and organised the appeal for the remainder.

The new schooner, which is almost identical on deck – only her square top deck-house doors distinguish her from the *Sir Winston Churchill* – was built by John Lewis & Sons of Aberdeen, and launched on 5th October, 1967. Below, alterations were made both to the trainee and after guard accommodation, as well as small improvements to the engine room, though she is also powered by two Perkins 135 hp diesels, driving variable pitch propellers. The trainees' heads were moved to the aft end of their half deck and the volunteer watch officers were shifted from their place near the counter on the *Sir Winston Churchill* to a cabin opposite the master, just below the companionway that leads up to the chart house.

The schooner office at Bosham, under Captain David Bromley-Martin, R.N., organises the ships and the supply of both trainees and after guard. A measure of the success of this organisation and the scheme itself is that both ships are over 99 per cent full through the year and have been throughout their lives. The cruises are limited to a fortnight to serve the needs of industry, and the schooners, which sail independently, cruise from one named port to another where, at the end of the fortnight, a new crew is picked up. The course between the two, which usually includes 'going foreign', depends on weather conditions.

The scheme is not restricted to boys and there have been up to a total of four girls' cruises in any one year on the two ships.

The cruises of the Sail Training Association ships differ from that of the Dulverton schooner, *Captain Scott*, in that the *Sir Winston Churchill* and *Malcolm Miller* are mostly at sea and do not have land expeditions built into their schedules. They cover some 30,000 miles a year with up to 36 cruises between them, taking some 1,200 young people to sea under training each year.

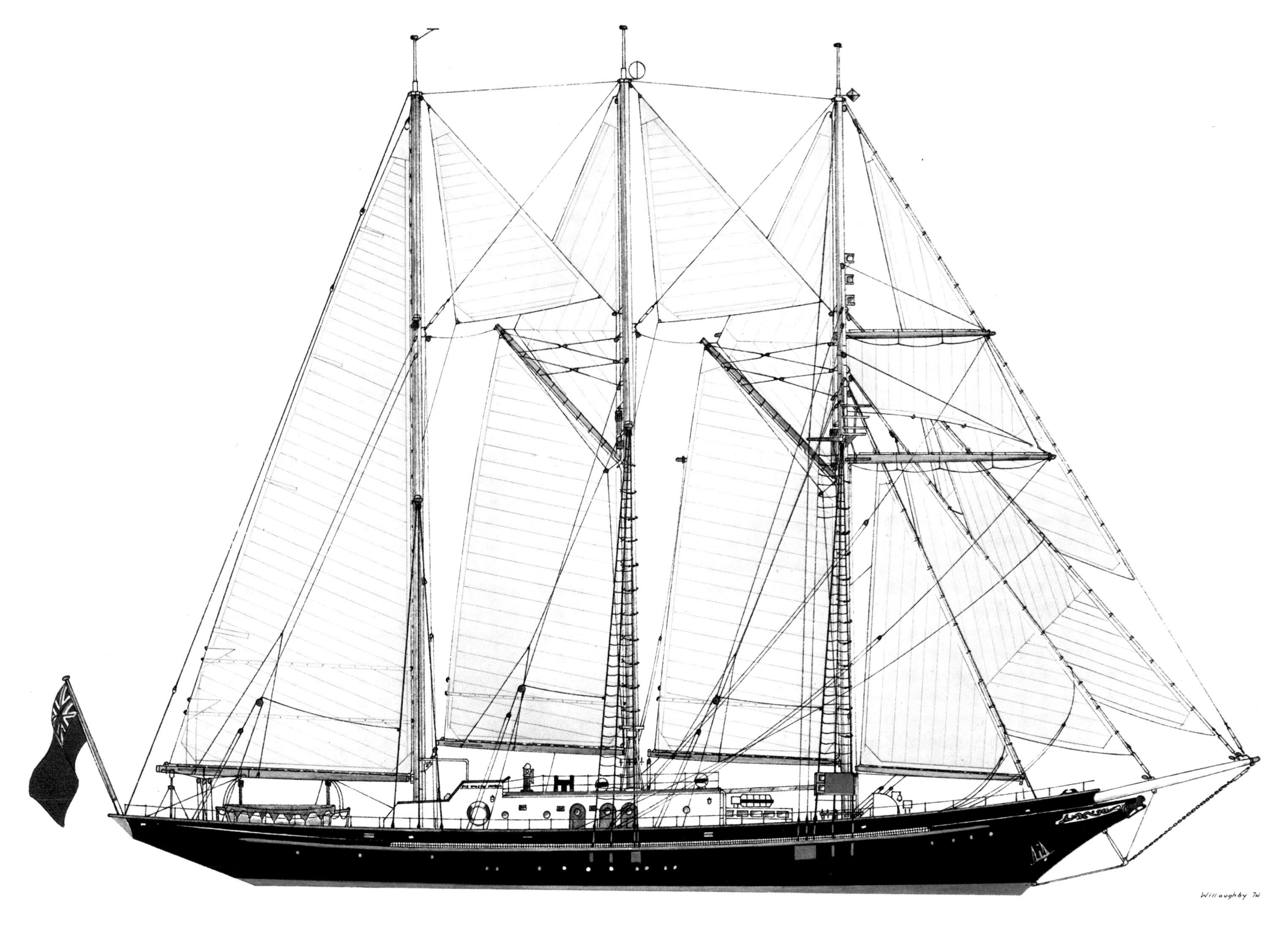

SIR WINSTON CHURCHILL

OWNERS: The Sail Training Association
DESIGNER: Camper & Nicholson
BUILDERS: *S.W.C.* Richard Dunston, *M.M.* John Lewis & Sons
PLACE OF BUILDING: *S.W.C.* Hessle, Hull, *M.M.* Aberdeen
YEAR OF BUILDING: *S.W.C.* 1966, *M.M.* 1967
BUILT FOR: Sail Training Association
RIG: 3-masted topsail schooner
CONSTRUCTION: Steel
L.O.A. (feet): 134·7
DRAUGHT (feet): 14·55
BEAM (feet): 25·0
TONNAGE: 299 T.M.
PRESENT NATIONALITY: British

The S.T.A.'s three masted topsail schooner,
Malcolm Miller, *close hauled, touching 12 knots.*

1. Breakfast, an important moment on the half deck: the Sail Training Association, as its pamphlet states, believes in addition to the classroom and the physical development of the body a third arm of education: 'The vital third, we maintain, is to teach young people to live and co-operate with their fellow human beings.'

2. Girls learn quickly: the two ships take 1,200 boy and girl trainees to sea each year, there being up to four girls-only cruises, which are well subscribed.

3. Course 230 degrees: each trainee has a spell at the wheel. The permanent crew never take the helm themselves.

4. Malcolm Miller *hard on the wind: the flying jib and gaff topsails are stowed.*

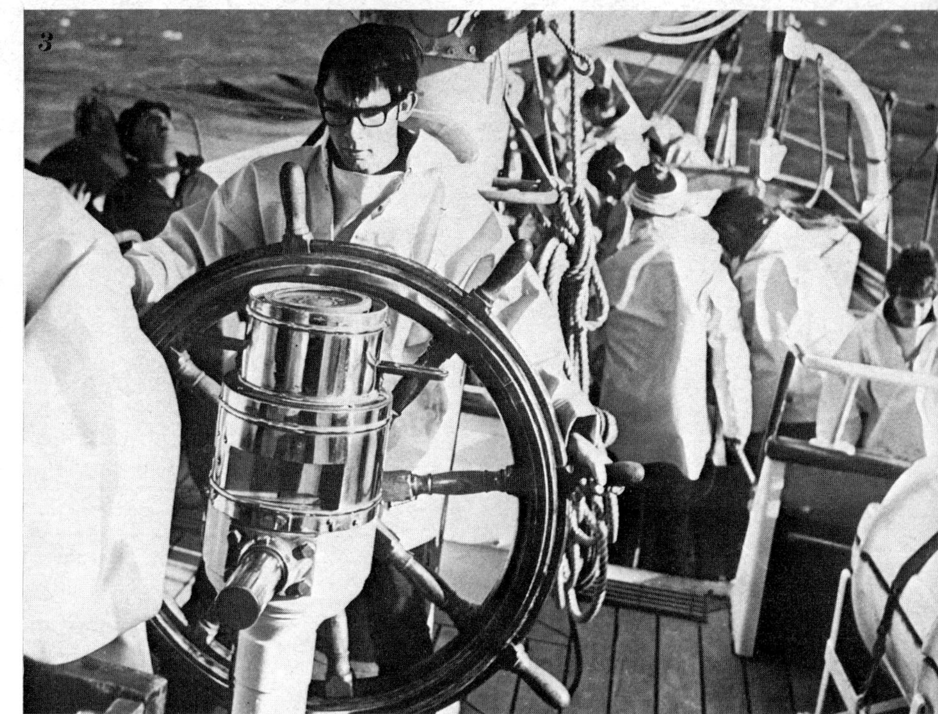

1. *Taking in sail, life harnesses clipped on.*
2. Malcolm Miller: *the square topped deckhouse doors identify her from the rounded ones of her sister ship,* Sir Winston Churchill.

TOVARISHCH

(ex *Gorch Fock*)

The 1,760 ton displacement steel barque, *Tovarishch II*, which was launched by Blohm & Voss on 3rd May, 1933 as *Gorch Fock I*, was the first of three barques built for the German Navy between 1933 and 1938. Her sisters were *Horst Wessel* (1936, now *Eagle*, see page 106), and *Albert Leo Schlageter* (now *Sagres II*, see page 118), completed in 1938. The Romanian barque *Mircea*, launched in 1919, is an exact sister ship and the German Navy's *Gorch Fock II* (1958) is similar, though 26 ft (8 m) longer over all.

Before the Second World War, *Gorch Fock I* cruised transatlantic in company with her sisters, her schedule designed to produce officers for the growing German Navy.

Named after a popular writer of sea stories who died at Jutland in the First World War, *Gorch Fock I* was built to replace the ill-fated German Navy training Jackass barque *Niobe* (ex Danish, four masted, fore and aft schooner *Morten Jensen*, built 1913). This vessel had sunk off the Fehmarn Belt Lightship after being struck by a sudden and vicious squall on 26th July, 1932. Sixty-nine were drowned, including 50 cadets, and only the Captain survived.

The tragedy resulted in the building of the new *Gorch Fock I*, building which was partially financed from funds raised by public subscription. There is some parallel between this and the Navy's latest barque *Gorch Fock II* which was commissioned on 17th December, 1958, after the loss of the *Pamir* the year before. Both disasters tested the German Navy's faith in sea training under sail.

All five of the barques had double topsails, single topgallants and royals. *Gorch Fock I*, unlike her sisters, had a single spanker, though *Horst Wessel*, on becoming *Eagle*, altered to this arrangement. *Sagres II* retained the double. All the barques were restricted in mast height as they were designed to pass through the Kiel Canal.

Below, the two large cadet flats in *Gorch Fock*'s 'tween deck are similar to those of her sisters. The petty officers and crew are accommodated forward while the officers are on the same deck aft, under the poop.

As designed, *Gorch Fock I* carried a total complement of 246, including 180 cadets. As *Tovarishch II* in the 1974 Tall Ships Race, from Copenhagen to Gdynia, the total number was 220, of whom 135 were cadets.

In 1942, in order to fit her better for war service, she was re-engined. However, her contribution ended when she was sunk off Stralsund in 1945, where she remained on the bottom until salvaged by the Russians in 1948.

In 1953, she was renamed *Tovarishch II*, the barque *Tovarishch I* having been originally the 2,301 ton displacement four masted, full-rigged ship *Lauriston*, built in Belfast in 1892 and sold to the Russians in 1913. As well as her name, *Tovarishch II* took over her predecessor's duties as a training ship for the Soviet Navy.

In 1974, as mentioned above, she took part in the Sail Training Association's Tall Ships Race from Copenhagen to Gdynia and also the *Britannia* review off Cowes.

OWNERS: Soviet Navy
PLACE OF BUILDING: Hamburg
RIG: 3-masted barque
DRAUGHT (feet): 17·0
BEAM (feet): 39·4
DESIGNER: Blohm & Voss
YEAR OF BUILDING: 1933
CONSTRUCTION: Steel
TONNAGE: 1,727 T.M.
BUILDERS: Blohm & Voss
BUILT FOR: German Navy
L.O.A. (feet): 203·7
PRESENT NATIONALITY: Russian

1. Tovarishch *taking part in the parade of sail, off Cowes, Isle of Wight, in 1974.*
2. *Making sail before passing the royal yacht* Britannia, *Cowes, 1974.*

KRUSENSTERN

(ex *Padua*)

The Russian, four masted steel barque *Krusenstern* of 3,545 tons gross, was built, as *Padua*, for the nitrate trade and the Hamburg shipowners Messrs F. Laeisz.

She was launched on 24th June, 1926 from the yard of her builders, J. C. Tecklenborg at Wesermunde. F. Laeisz already owned five, four masted barques and two ships, including the four masted barque *Pamir* (see page 32), when he ordered *Padua*. The Flying P Line certainly believed that sail could pay in the face of steam.

Padua was a 'three island' barque, a design favoured by Laeisz, for in addition to the poop and forecastle, the midship deck-house was brought out to the ship's side, forming the centre island. Although she was basically designed for cargo carrying, cadets were provided for at the outset, and *Padua* carried 40 boys under instruction. They were taught all aspects of practical seamanship.

On her maiden voyage from Hamburg, she made Talcahualo, Chile in 87 days. Tragedy struck in 1930 – rounding Cape Horn in severe weather, she lost four men overboard.

However, *Padua* was a consistently good performer, and when she later moved to the Australian grain trade, she managed Hamburg to Port Lincoln, Australia in 67 days. According to Laeisz's figures, *Padua* achieved 351 nautical miles, from noon to noon, on 28th December, 1933. In 11 consecutive days on the same trip she covered 3,123 nautical miles, which made her champion of the company's Flying P fleet. Incidentally, W. L. A. Derby in *Tall Ships Pass*, records that *Herzogin Cecilie* (see page 24), managed 365 nautical miles in 23½ hours on 5th December, 1930.

Sailing ships, however, suffered from the depression in world trade and *Padua* was laid up in 1932. Her rest was short lived and she was soon back at sea. In 1939, she recorded a run of 96 days, Australia to Europe.

At the end of the war, the barque was in Flensburg, and in January 1946, she was taken over by the Russians who later also obtained the Krupp-built, four masted, steel barque *Kommodore Johnsen* (ex *Magdalene Vinnen*, 1921, Hamburg) which they renamed *Sedov*. *Padua* was renamed *Krusenstern* by her new owners, after the famous Russian navigator and hydrographer, Admiral Ivan Krusenstern (1770–1846) who had advocated the advantages of a direct trade route between Russia and China via Cape Horn and the Cape of Good Hope. He was commissioned in 1803 to survey the route and became the first Russian sailor to circumnavigate the world in 1806. *Sedov* took her name from G. L. Sedov, the Russian Polar explorer who died in the Arctic attempting the Pole in 1913.

Krusenstern is the training ship of the Fishery Board of the U.S.S.R. and carries a complement of 236, made up of 26 officers, 50 petty officers and crew and 160 cadets. She is now equipped with a Russian diesel engine as an auxiliary, and is painted black with gun ports, in the manner of a Blackwall frigate.

In 1974, she entered the Tall Ships Race, from Gdynia to Copenhagen with the barque *Tovarishch II* and won Class A. In the same year, she made a particularly fine sight in the Royal Yacht *Britannia* review off the Isle of Wight, sailing through Cowes Roads under all plain sail.

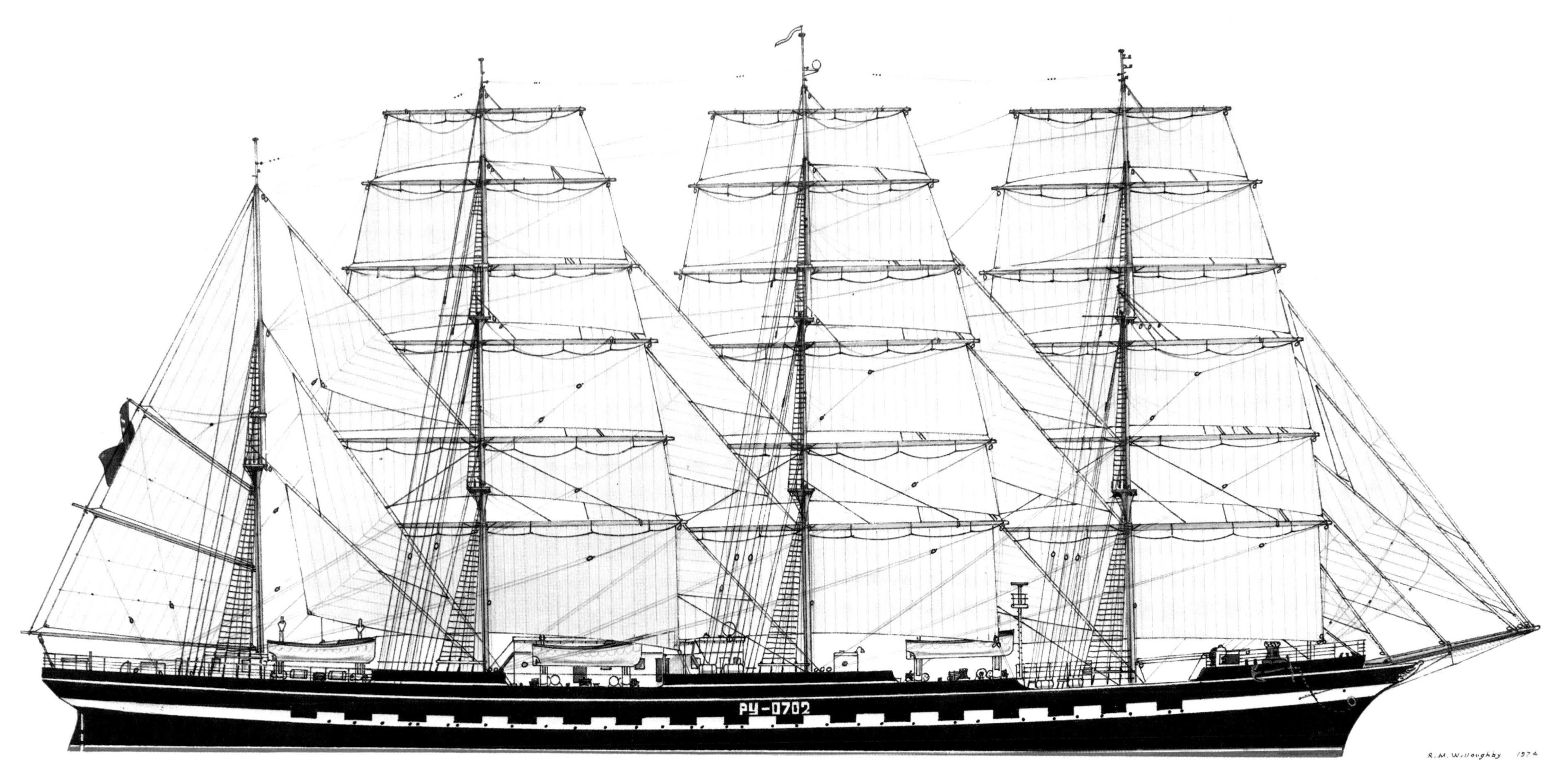

KRUSENSTERN

OWNERS: Fishery Board of the U.S.S.R.
PLACE OF BUILDING: Wesermunde
RIG: 4-masted barque
DRAUGHT (feet): 22·64
BEAM (feet): 42·62

DESIGNER: J. C. Tecklenborg
YEAR OF BUILDING: 1926
CONSTRUCTION: Steel
TONNAGE: 3,185 T.M.

BUILDERS: J. C. Tecklenborg
BUILT FOR: Ferdinand Laeisz
L.O.A. (feet): 342·0
PRESENT NATIONALITY: Russian

1

1. The midday sight.
2. Leadership, example, and the ability to live and work together are learnt aboard.
3. Passing the royal yacht Britannia *in the parade of sail, Cowes Roads, after the 1974 Tall Ships Race.*

3

EENDRACHT

H.R.H. Prince Bernard of the Netherlands launched the steel auxiliary schooner *Eendracht* on the 1st June, 1974. The Prince is the Patron of the owners, Stichting 'Het Zeilend Zeeschip'. The name *Eendracht* means 'concord'. A predecessor was the famous Admiral Tromp's flagship in the middle of the 17th century, and the name is continued in the Royal Dutch Navy to this day. The aim of the schooner and her organisation is very similar to that of the British Sail Training Association's two three masters *Sir Winston Churchill* and *Malcolm Miller*. Indeed, Kaes van Dam, Holland's representative on the Sail Training Association's International Race Committee, is the Chairman of the Nationale Vereniging 'Het Zeilend Zeeschip' which started the project and also chairs the Stichting (i.e. foundation) of the same name, formed for the purpose of running her.

The *Eendracht* was designed by W. de Vries Lentsch and built by the Amsterdam shipyard of Commenga. A two master, she carries a bermudan main, gaff foresail, with topsail, squaresail and raffee, and has a raised poop, with the wheel-house on the fore end. There are two steering positions, one inside and the other, the main one, aft. The five permanent crew are accommodated under the poop in two double cabins, with a single for the skipper. The sunk deck-house amidships contains the mess, and the ship's company eats together, officers included. Twelve of the 32 trainees live in the forecastle, and the remainder are divided between the three, four and the single six berth cabins. This layout below enables a mixed crew to be carried.

The ship is all electric, including cooking, with power provided by two generators. Under power the schooner is propelled by a 400 hp GM diesel, driving a single screw. Special attention has been paid to silencing the engine room and this has been most effective, overcoming the usual problems of alien noise aboard a sailing vessel.

The trainees are drawn from all over Holland and are welcome from other parts of the world. The junior age group is 16–26 and the two older ranges start at 16 and 21 respectively, up to any age fit enough to play a part aboard. The cruises are of different lengths, the usual being seven, ten or fourteen days. They start and finish from a variety of places in European waters, exchange being made by bus or charter flight.

Eendracht leaves northern waters for the Mediterranean in mid-October, returning to Holland at the end of the next year. Several cruises of 14 days, served by charter flights to Gibraltar, for the age group 16 and over, are organised for this period. This enables the schooner's programme to be extended, leaving only five weeks for the refit. In comparison, the Sail Training Association's schooners, *Sir Winston Churchill* and *Malcolm Miller*, spend over two months out of commission while refitting in the Port of London. *Eendracht* invariably takes part in the Tall Ship Races.

An unusual feature of the schooner is that the master is a volunteer and is not paid as such. There is a small roster of captains, the senior master being Commander Juta, who has an extra master's ticket. So far they have all been ex Royal Netherland's Navy officers. The remaining four of the permanent crew are paid at merchant navy rates – they are a mate with a second mate's ticket, engineer, boatswain and cook. This is the same complement as the British schooners and, in a similar way, *Eendracht* has three volunteer watch officers.

The Dutch merchant navy regulations insist that any vessel of *Eendracht*'s size running to such a schedule, must have a doctor aboard if she is carrying trainees. The Dutch Medical Council has, therefore, organised and paid a young doctor for each cruise. His period aboard counts as hospital time and the arrangement has proved most successful.

The schooner's finish is exceptional and she is smartly kept, although in commission for almost the whole year. The flexibility of her layout enables the organisation to carry out a useful adventure training schedule and yet provide an opportunity for cruises that can help subsidise the main youth training purpose of the ship.

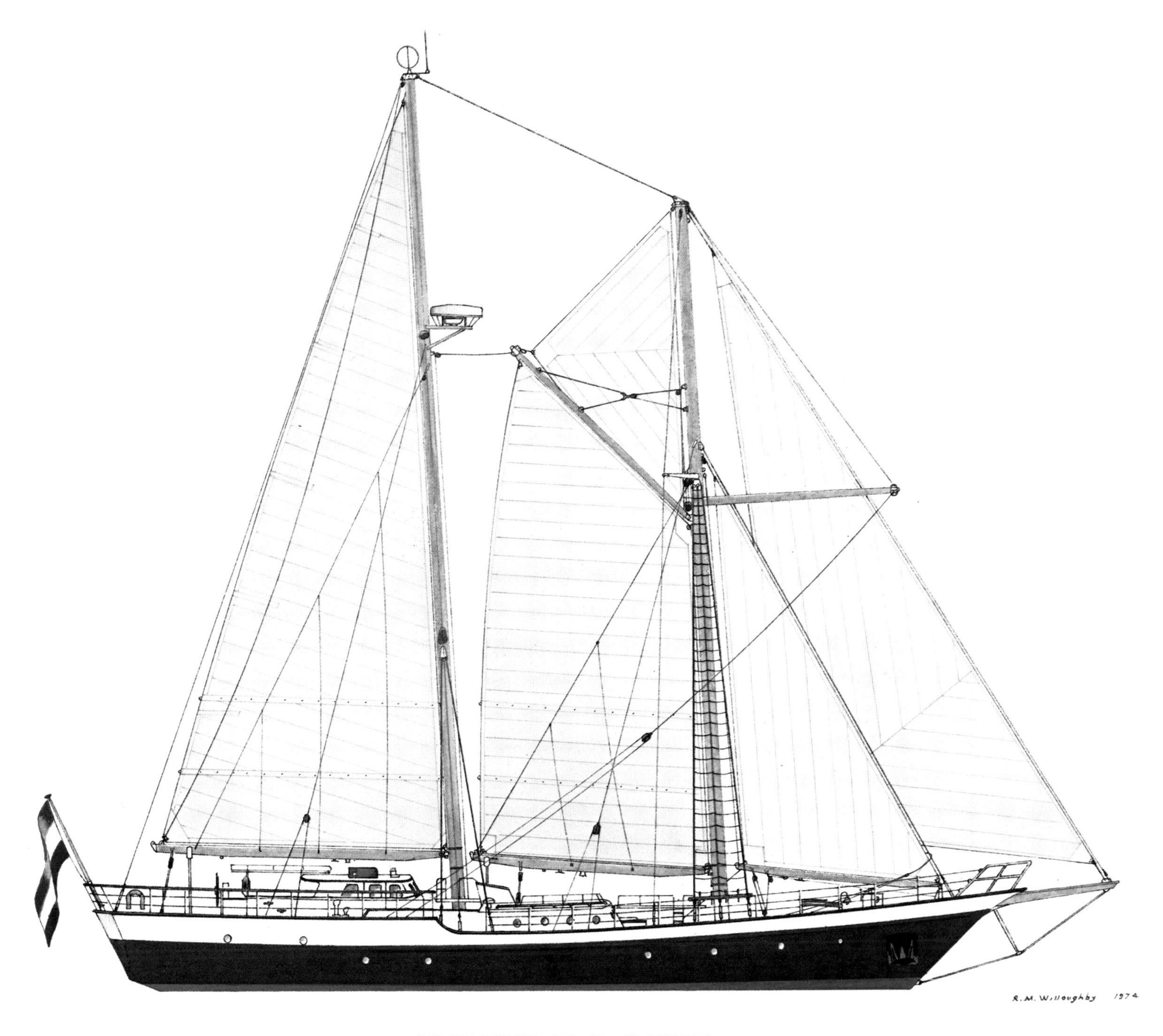

EENDRACHT

OWNERS: Nationale Vereniging 'Het Zeiland Zeeship'
DESIGNER: W. de Vries Lentsch
BUILDERS: Commenga
PLACE OF BUILDING: Amsterdam
YEAR OF BUILDING: 1974
BUILT FOR: Nationale Vereniging 'Het Zeilend Zeeship'
RIG: Schooner
CONSTRUCTION: Steel
L.O.A. (feet): 107·1
DRAUGHT (feet): 10·77
BEAM (feet): 26·12
TONNAGE: 226 T.M.
PRESENT NATIONALITY: Dutch

1. Going aloft to set the course.
2. Close hauled: the 32 trainees are drawn from all over Holland and welcome volunteers from other parts of the world. The majority of the Cruises start from Gibraltar in the winter. cruises start from a variety of European ports during the spring and summer, but begin from Gibraltar in the winter.
3. A Chartroom conference: the schooner carries a permanent crew of five, and the Master is drawn from a roster of ex-Royal Dutch Navy Officers. A doctor is also carried, provided by the Dutch Medical Council and there are three volunteer Watch Officers, signed on for each cruise.

EAGLE

The 1,800 ton displacement steel barque *Eagle* was built by Blohm & Voss at Hamburg in 1936 and named *Horst Wessel*, after a young Nazi leader. Hitler attended her launch. The *Horst Wessel* was one of three sail training barques built for the German Navy between 1933 and 1938.

The first of these, *Gorch Fock I*, now the Russian training barque *Tovarishch II* (see page 96), was launched in 1933. *Horst Wessel* was next and the barque *Albert Leo Schlageter*, now the Portuguese Naval training barque *Sagres II* (see page 118) was the third, having been completed in 1938. The Romanian barque *Mircea II* (see page 138), built in 1939, is a near sister.

After commissioning, and in the few years of peace that were left, *Horst Wessel* cruised to the Canaries and West Indies, but returned to the comparative safety of the Baltic before war broke out. After the outbreak, she transported men and supplies to and from East Prussia. Her log claims that she fired at Allied aircraft, and she is credited with downing three Russian planes, according to the ship's historian, William I. Norton, in his book *Eagle Ventures*. Approaching Kiel during the last few days of the war she saw, but escaped, the crushing air raid that the city experienced. Her Captain altered course for Flensburg and the ship eventually berthed at Bremerhaven at the end of the war.

In the meantime, the United States Coast Guard had enjoyed the loan of the full-rigged ship *Danmark* (see page 52) which had spent the war years training American cadets. Captain Knud Hansen of *Danmark* suggested that *Horst Wessel* would make a good replacement for the Danish ship, and Commander Gordon P. McGowan, who had been with Hansen aboard *Danmark*, was sent over to retrieve *Horst Wessel*. He found her, her grey paint rusted and scarred, in a bombed out Bremerhaven shipyard. A scratch crew refitted her with difficulty, removing the swastika from the claws of the oversized eagle figurehead that decorated her bow, and replacing it with the Coast Guard emblem. The barque arrived in New London, her new base, in the summer of 1946, having been caught in a hurricane on the way across the Atlantic. On her arrival, she was renamed *Eagle*, following a long line of United States Revenue cutters. The oversized figurehead was replaced for a period of several years with the smaller and more graceful one which had belonged to the Coast Guard Academy's training barque *Chase* (1878 to 1907, see page 30). In the late 1960s a fibreglass moulding of the original eagle figurehead was reinstalled. Her rig, however, has remained virtually untouched, except for the mizzen, which was rearranged to take a single rather than a double spanker.

The Americans kept the same accommodation below, there being two full length steel decks with a platform deck below these with a raised poop and forecastle. The 180 cadets live on the vinyl covered 'tween or second deck in two flats, sleeping in hammocks, with the 46 permanent crew forward of them. The officers are accommodated in the traditional way aft. Thus the ship carries three times as many cadets as a modern Coast Guard cutter.

In June of every year since 1948, the cadets of the first and third classes aboard *Eagle* and her accompanying cutters depart on a two and a half month cruise to Europe or the Caribbean. On her return to New London in mid-August, the second and 'swab' classes make a short cruise in the western Atlantic. On a cadet's first cruise he stands watches and performs duties in the same way as an enlisted man on an ordinary Coast Guard cutter, but with the additional job of locating and understanding the 154 leads required to sail the vessel.

The upper classmen undertake jobs normally carried out by officers or senior petty officers. Alan Villiers asked Captain Carl Bowman of *Eagle* why the Coast Guard sticks to an anachronism like the training ship. The Captain replied, 'We can see our boys here all the time, and we get a pretty good idea of what they are made of before the voyage is over.'

Eagle's auxiliary is an eight cylinder, four stroke direct reversing diesel, which drives the propeller, through a single reduction gear with a 2·9:1 ratio. At 580 rpm, the engine is rated at 750 hp. A special 'sailing clutch' disengages the shaft from the reduction gear, allowing the propeller to turn freely while the vessel is under canvas. Three 75 kw generators provide electrical power and the barque carries 56,000 gallons of fresh water, which can be reinforced by an evaporator at the rate of 2,500 gallons a day; 23,000 gallons of fuel in five tanks are also provided.

Though designed as a sailing vessel, *Eagle* is one of the most sophisticated of the six United States Revenue Service and Coast Guard vessels to bear the name. The first *Eagle* was an 187 ton brig, built in 1798. The fifth *Eagle* was much involved in upholding the 'noble experiment' of prohibition.

Training under sail aboard *Eagle* certainly reflects the Coast Guard Academy's tradition:

'To graduate young men with sound bodies, stout hearts, and alert minds, with a liking for the sea and its lore, and with a high sense of honor, loyalty and obedience which goes with trained initiative and leadership; well-grounded in seamanship, the sciences and the amenities; and a strong resolve to be worthy of the traditions of the commissioned officers in the U.S. Coast Guard in the service of their country and humanity.'

EAGLE

OWNERS: U.S. Coast Guard Academy
DESIGNER: Blohm & Voss
BUILDERS: Blohm & Voss
PLACE OF BUILDING: Hamburg
YEAR OF BUILDING: 1936
BUILT FOR: German Navy
RIG: 3-masted barque
CONSTRUCTION: Steel
L.O.A. (feet): 265·3
DRAUGHT (feet): 15·96
BEAM (feet): 38·75
TONNAGE: 1,634 Disp.
PRESENT NATIONALITY: American

1. In a big swell off Bermuda: built as Horst Wessel *for the German Navy in 1936, she was taken over by the U.S. Coast Guard in the summer of 1946.*
2. A cadet at the helm: the double wheel enables assistance to be given in heavy weather.
Eagle *carries 180 cadets and 46 permanent crew.*

2

1. Feeding Elmer, Eagle's *auxiliary engine.*
2. Checking away slowly.
3. Walking away with it: as Captain Bowman remarked to Alan Villiers, 'We can see our boys here all the time, and we get a pretty good idea what they are made of before the voyage is over.'

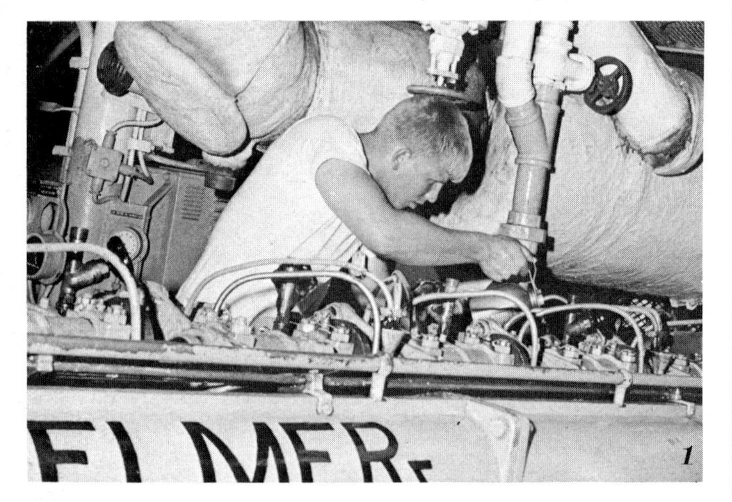

1

2

3

1. Fancy work: making a rope fender.
2. Fair weather off the eastern seaboard of the United States.
3. Eagle *spirit: 'To graduate young men with sound bodies, stout hearts and alert minds, with a liking for the sea and its lore . . .'*

WESTWARD

The only full-time civilian offshore sail training ship in the United States is the 220 ton, steel research schooner *Westward*. She is owned by S.E.A., the Sea Education Association, Inc., which operates her the whole year round.

Westward provides the sea training for 126 students each year, in intensive 12 week courses. The six weeks ashore, designed for 270 a year, takes place at Woodshole, Cape Cod, the home of the Woodshole Oceanographic Institution, one of the world's leading centres for marine research. In addition, the little seaport supports the National Marine Fisheries Service and the Marine Biological Laboratory. Woodshole, therefore, is a unique centre for both the marine scientist and the seaman.

Students may take the shore course by itself and many do, thus obtaining eight credits from Boston University. A full course including sea training aboard *Westward*, entitles them to 16 academic credits.

Westward, therefore, is designed to act as the practical arm of the programme, the students having obtained theoretical understanding of the oceans while ashore. S.E.A. is convinced that a sailing ship provides the most effective means for this form of education, and experience shows that six weeks at sea are needed to develop the necessary understanding and skills in the average student.

This requirement has determined the layout of *Westward*. Her rig, that of a bermudan staysail schooner with fore course and raffee, is simplicity itself compared with other training ships. It is enough, however, to give the trainees experience of sail without the need for continuous and arduous attention, and means that the research work and practical and scientific experiments can be carried out under sail. But for all this, she carries 7,000 sq ft (650 sq m) of canvas.

The schooner was built in 1961 by Abeking & Rusmussen, Germany, to Lloyd's highest classification. She was designed by Eldridge & McInnis of Boston as a private yacht, on the lines of *Yankee*, for Drayton Cochrane of Oyster Bay, New York. He sailed her extensively, visiting both the Atlantic and Pacific, as well as the Mediterranean – before selling her to the Oceanic Foundation at Makapuu, Hawaii where she undertook scientific investigations until bought by S.E.A. in 1971.

The hull is divided into 11 watertight compartments. Each apprentice, as he or she is called – and there are 21 aboard – has his own curtained-off bunk space 76in (193cm) × 30in (76cm) × 30in (76cm). The regular professional staff of eight includes the captain, staff scientist, three watch officers, an engineer, a steward and a doctor, when appropriate. There is also accommodation for as many as three visiting scientists.

The captain normally holds a master's ticket and is on the staff of Boston University. The staff scientist looks after the programme of research work and the accommodation of the visiting scientists, and is also responsible for the scientific instruction aboard. The steward acts as purser, and the engineer is responsible for the engine room and all mechanical services.

Apprentices are divided into three watches with a watch officer in charge of each. The watch officer is responsible for the welfare, discipline and education of those in his watch. The system worked is four hours on and eight hours off and, while at sea, each watch is responsible for the safe navigation of the ship, as well as the collection and processing of oceanographic data. In addition to this, normal duties in the galley, engine room and laboratory, are carried out.

In the six weeks aboard, an average of 10 days per course is spent in harbour. This is usually given over to maintenance and the writing up of oceanographic field work. In this way, *Westward* spends 250 days at sea and in this time covers some 20,000 miles.

As can be seen, the course is a balanced one with both seamanship and oceanography built into it. During the three weeks aboard, each apprentice has to perform the duties of a junior officer, taking the job of boatswain, navigator, steward, engineer, scientist of the watch and junior watch officer in turn. The idea is learning by doing.

Westward is well organised for her oceanographic work, carrying winches for sampling depths up to 2,734 fathoms (5,000 m), a small laboratory, deep sea echo sounder and bathythermographs, plankton nets and various water sampling devices. Additional equipment is brought aboard for special programmes.

As S.E.A. explains, 'An expedition aboard *Westward* clarifies the apprentices' understanding of the field of oceanography and confirms or ends their interest in it. An apprentice's time off is uncluttered with much of the trivia of modern life. He has time to marvel at the changing aspects of weather, at the effortless flight of the sea bird, at the majesty of the ocean. He has time to read, to write, to listen to a shipmate, to sort out his own thoughts.' This last sentence is a very good description of the 'sea change' that adventure training can work.

Westward is a pioneer in this form of training under sail, wedded as she is to the exploration of the ocean. S.E.A. hope to acquire more ships of her type and so provide more places for the growing number who wish to take advantage of this unique course.

OWNERS: Sea Education Association
DESIGNER: Eldridge & McInnis (Boston)
BUILDERS: Abeking & Rusmussen
PLACE OF BUILDING: Germany
YEAR OF BUILDING: 1961
BUILT FOR: Drayton Cochrane
RIG: Bermudan staysail schooner
CONSTRUCTION: Steel
L.O.A. (feet): 135
DRAUGHT (feet): 12·5
BEAM (feet): 21·5
TONNAGE: 220 Disp.
PRESENT NATIONALITY: American

1. Inspecting, and learning to understand, icebergs in the north-western Atlantic: She continually monitors the sea environment, measuring pollution created by man, as well as carrying out plankton surveys, studies of marine temperature and chemistry, marine mammal distribution and productivity.
2. Sort this lot out: a cruise lasts six weeks and she spends an average of 250 days a year at sea.

LA BELLE POULE & L'ETOILE

The two wooden 227 ton displacement French naval training schooners, *La Belle Poule* and *L'Etoile* are nearly identical, and rank among the most attractive vessels afloat today.

They were launched in January, 1932, to the order of the French government for the Marine Nationale Ecole Naval, at Brest. Both ships were built at Fecamp by Chantiers Naval de Normandie which had a tradition of building fine sailing ships. The lines of *La Belle Poule* and *L'Etoile* came from the Paimpol schooners that used to fish for cod in Icelandic waters.

These white hulled, two masters, with their long bow sprits, distinctive martingale and deep roller reefing square topsail are instantly recognisable. Before the Second World War, coasting schooners of their type were common, working their way up and down the Channel and in the ports of the Brittany coast.

The two naval schooners sail in company, cruising in the Channel and off the West Coast of France during the summer months, training naval regular and reserve officers. Each ship carries a permanent crew of 20, which includes three officers, and 30 cadets. Both invariably take part in the Tall Ships Races.

According to Otmar Schauffelen, in *Great Sailing Ships*, the name *La Belle Poule* was first used in the French Navy after a pirate ship of that name had rendered such service to the country that Louis XV decreed that the name should be preserved. Because of this, three frigates before the present schooner have borne the name. One of these brought back Napoleon's exhumed body to France in 1840. *L'Etoile* is one of the favourite names of the French Navy, having been borne by 15 ships, including this one, since 1622.

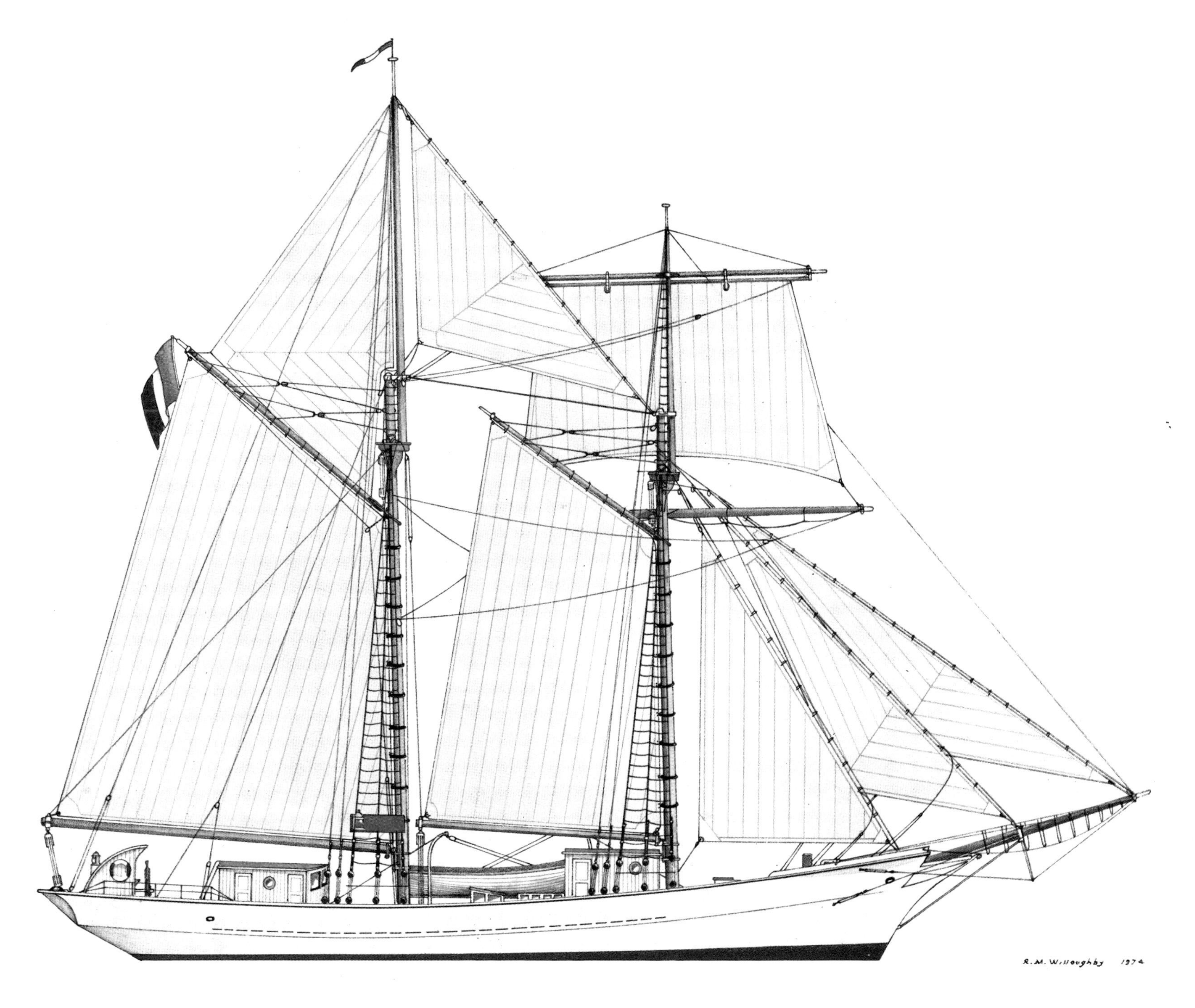

LA BELLE POULE

OWNERS: French Navy
DESIGNER: Chantiers Naval de Normandie
BUILDERS: Chantiers Naval de Normandie
PLACE OF BUILDING: Fécamp
YEAR OF BUILDING: 1932
BUILT FOR: Marine Nationale Ecole Naval, French Navy
RIG: Topsail schooner
CONSTRUCTION: Wood
L.O.A. (feet): 99·27
DRAUGHT (feet): 11·51
BEAM (feet): 24·28
TONNAGE: 227 T.M.
PRESENT NATIONALITY: French

The twins close hauled: they are perhaps the best looking of all the training schooners.

SAGRES II

The three masted steel barque *Sagres II*, is the Portuguese Navy's second training ship of that name, both called after the 'Infanta's Town' near Cape St. Vincent, where Prince Henry the Navigator often lived, and oversaw the Naval arsenal. It was his court at Sagres that had most responsibility for geographical study and practical exploration in the first half of the 15th century, and he died there in 1460. His is the bust on the figurehead of *Sagres II*.

Both ships were originally German – the first, a steel full-rigger, began life as *Rickmer Rickmers* of 1,980 tons, built by the Rickmers family in their Bremerhaven yard and launched in 1896; the second began life as the *Albert Leo Schlageter* of 1,869 tons, built by Blohm & Voss of Hamburg and launched in 1937.

From 1896 to 1912, *Rickmer Rickmers* plied between Portugal and the Far East with general cargo. In the latter year, she was bought by C. Kraabbenhoft of Hamburg, rechristened *Max* and adapted for the nitrate trade. Seized by the Portuguese in 1916 and renamed *Flores*, she became a carrier of war material; then disappeared from the register until 1924 when she reappeared as *Sagres I*, a training ship for the navy. In 1961 she became a stationary seamen's school in the river Tagus, under the name of *Santo Andre*.

Albert Leo Schlageter took over her predecessor's role in 1961 when, as *Guanabara*, the Brazilian Navy's training ship, she had been bought and rechristened *Sagres II*. Between her launching and this event, she had made very few voyages, having been damaged by mines during the Second World War and passed by the United States to Brazil in 1948. However, during the past 15 years, she has overcome these vicissitudes, taking part, for example, in the Tall Ships Race in 1964 with Alan Villiers aboard.

OWNERS: Portuguese Navy
PLACE OF BUILDING: Hamburg
RIG: 3-masted barque
DRAUGHT (feet): 17·19
BEAM (feet): 39·44
DESIGNER: Blohm & Voss
YEAR OF BUILDING: 1937
CONSTRUCTION: Steel
TONNAGE: 1,784 T.M.
BUILDERS: Blohm & Voss
BUILT FOR: German Navy
L.O.A. (feet): 267·7
PRESENT NATIONALITY: Portuguese

Under full sail on the Tagus River, Portugal. The appearance of Sagres II *is very distinctive with large red Portuguese crosses on her squaresails.* Sagres I *used to display one also on her spanker.*

LIBERTAD

The Argentine Navy's full-rigged ship *Libertad* of 3,025 tons displacement is the largest sail training ship in the world today. She was launched on 30th June, 1956 from the Argentine State Shipyard, Astilleros y Fabricas Navales Del Estado, and commissioned in 1960.

The ship has an imposing appearance, being a full-rigger with a flush deck. This is relieved by a modern bridge between the fore and main mast. She is powered by two 1,200 hp Sulzer diesels, connected to a single shaft, which gives her a speed under power of over 13 knots. The funnel is between her main and mizzen masts.

Libertad has a total complement of 520, which includes 150 watch keeping and engineering midshipmen. After four years at the Naval College, every midshipman has to undergo a training cruise around the world aboard her. A combination of hard work, discipline and adventure is the keynote, for on arrival back in Buenos Aires, they have to take their final passing out examination.

She made her first cruise, east via Capetown, in 1963. In 1966 she covered the 1,741·4 miles from Cape Rice, Canada to Burcey Island, Ireland in six days, 21 hours, at an average speed of 10·5 knots. At times she managed an average of 18 knots and won the Sail Training Association's Boston Tea Pot trophy as a result.

The ship's stern is embellished with the arms of the Argentine Navy, while a female figure decorates the bow.

Libertad replaced *President Sarmiento*, a full-rigger with twin funnels built on the lines of a 19th century corvette. This training ship was built by Cammell Laird at Birkenhead in 1897 and named after President Domingo Faustino Sarmiento, who did much to build up the Argentine Navy. *President Sarmiento* is now maintained as a stationary school ship at the Argentine Naval Academy.

OWNERS: Argentine Navy
DESIGNER: Argentine State Shipyard
BUILDERS: Argentine State Shipyard
PLACE OF BUILDING: Rio Santiago
YEAR OF BUILDING: 1956
BUILT FOR: Argentine Navy
RIG: Ship
CONSTRUCTION: Steel
L.O.A. (feet): 298·2
DRAUGHT (feet): 22·88
BEAM (feet): 45·0
TONNAGE: 2,587 T.M.
PRESENT NATIONALITY: Argentine

1. Dressed over all: the full rigged Libertad *was built in 1956 by the Argentine State Shipyard.*
2. In Sydney Cove attending the Captain Cook Bicentenary Celebrations in April 1970.

2

NIPPON MARU

(and her sister ship *Kaiwo Maru*)

The Japanese Government has always believed in the virtues of sea training under sail, and the four masted sister barques *Nippon Maru* and *Kaiwo Maru* are continuing evidence of this. They were preceded by the four masted barque *Tasei Maru*, which was built in 1904, and was the prototype for her two successors which were built by Kawasaki's yard at Kobe in 1930. *Nippon Maru* and *Kaiwo Maru* are very nearly identical sisters though the latter has less draft and so the gross tonnages vary, *Nippon Maru* being 2,257·72 tons while her sister ship is 2,250·62.

Both ships were designed at the outset as training ships and the maximum use has been made of their length in order to produce as much accommodation and classroom space as possible. The poop extends nearly to the fo'c'sle, leaving little of the main deck open before reaching the short fo'c'sle. The bridge is forward of the main mast and it is from here that the ship is steered when under power, leaving the twin wheeled position abaft the mizzen for control under sail.

The cadets' accommodation is unusual, particularly at the time of building, for it consists of 15 eight berth cabins, providing accommodation for 120 in all. These cabins are arranged along the ship's sides, leaving a long central flat that is used for instruction and recreation. Both ships carry 76 permanent crew, which includes ship's officers and classroom instructors.

The engine room has been designed to assist in the training programme, being sufficiently large to take cadets as well as auxiliary machinery and the two 600 hp diesel engines that are the main propulsion units. The funnel between the main and mizzen masts serves as main engine and auxiliary exhausts and trunking for other services.

Both ships possess water ballast tanks in their double bottoms and carry 640 tons of copper and 130 tons of iron set in concrete.

The rig, which is conventional, uses single pole masts. The rigging and sails were from Ramage & Ferguson of Leith, Scotland. Sail handling is assisted by six capstans.

The ships were built as training vessels to serve all the public mercantile marine schools of Japan. Cadets have to do up to three years in a shore establishment before spending a year at sea, six months of which is on board one of these two vessels. The remaining period, before qualification as a second mate, is completed on one of the Ministry's five steam or motor vessels. The ships, therefore, play a very significant part in the efficient operation of the Japanese merchant service and their smart upkeep is indicative of this.

During the Second World War, both barques were derigged, their yards and canvas being sent ashore before they passed to the Ministry of Transport. They continued training under power, although they were used for transporting coal and cargo mainly on the Japanese inland sea.

After the war, *Nippon Maru* resumed training in June, 1952 and *Kaiwo Maru* in 1955. In the post-war pattern of cruises, the barques, which cruise independently, begin their season in May and include an ocean-going voyage to the west coast of America or the Hawaiian Islands. The name *Kaiwo* is the equivalent of *Sovereign of the Seas*. *Nippon*, of course, refers to Japan.

NIPPON MARU

OWNERS: Ministry of Transport	DESIGNER: Kawasaki Yard	BUILDERS: Kawasaki Yard
PLACE OF BUILDING: Kobe	YEAR OF BUILDING: 1930	BUILT FOR: Ministry of Transport
RIG: 4-masted barque	CONSTRUCTION: Steel	L.O.A. (feet): 306·9
DRAUGHT (feet): *N.M.* 22·6, *K.M.* 21·9 BEAM (feet): 42·6	TONNAGE: *N.M.* 2,257·7 gross, *K.M.* 2,250 gross	PRESENT NATIONALITY: Japanese

1. Observing the altitude of the sun: each ship carries 120 cadets and 76 permanent crew.
2. Boat drill in mid-Pacific: Nippon Maru *is in the background.*
3. Wearing ship: Nippon Maru *was built to serve all the public maritime schools of Japan. Cadets spend three-years in shore establishments before spending a year at sea.*

GORCH FOCK II

Gorch Fock II, launched on 23rd August, 1958, is the last of a line of German Navy training barques that started in 1933 with *Gorch Fock I* (now the Russian *Tovarishch*). The others, in order of building, were *Horst Wessel* (now the U.S. Coast Guard *Eagle*, see page 106), built in 1936; *Albert Leo Schlageter* (now the Portuguese *Sagres II*, see page 118) launched in 1938; and *Mircea*, built by Blohm & Voss, Hamburg, for the Romanian Navy in 1938 (see page 138).

The launching of *Gorch Fock II* indicated the German Navy's belief and, indeed, Germany's confidence in the value of sail training, as this took place after the tragic loss of the four masted barque *Pamir* (see page 32) on 21st September, 1957, 600 miles south-west of the Azores. *Pamir* was capsized by hurricane Carrie with a loss of 80 men, of whom 52 were cadets.

As was to be expected, stability and safety were the keynotes in the design of the new ship. Indeed, the first *Gorch Fock*, which was of the same design, was built as a replacement for the three masted Jackass barque *Niobe* that foundered after being capsized off the Fehmarn Light Vessel on 26th July, 1932 with a loss of 69 lives. The Blohm & Voss training ship design was well tested with the five subsequent sister ships, as each had proved so successful in the 24 years since the first *Gorch Fock* was launched. It is doubtful whether any sailing vessel has had so much attention paid to her stability and safety.

The Federal German Navy's training barque carries a total complement of 269, including 200 cadets. They are accommodated, as in the older barques, on two flats on the 'tween deck.

Her rig is similar to the others, double topsails, single topgallant and royals, though she, like *Sagres*, carries a double spanker.

Gorch Fock II has taken part in nearly all the major Tall Ship Races and probably is the most successful competitor in Class A since the series began. Her smartness is legendary. From 1964 to 1968, she was commanded by Captain Hans Engel. He was the principal organiser of 'Operation Sail', Kiel, in 1964 and is the Sail Training Association's German representative.

GORCH FOCK II

OWNERS: Federal German Navy
DESIGNER: Blohm & Voss
BUILDERS: Blohm & Voss

PLACE OF BUILDING: Hamburg
YEAR OF BUILDING: 1958
BUILT FOR: Federal German Navy

RIG: 3-masted barque
CONSTRUCTION: Steel
L.O.A. (feet): 265·8

DRAUGHT (feet): 15·58
BEAM (feet): 39·21
TONNAGE: 1,727 T.M.
PRESENT NATIONALITY: West German

1. The eyes of the ship: Gorch Fock II *is the last of the Blohm and Voss training barques.*
2. Full and bye: she can set 21,000 sq. ft. (6,384 sq. m.) of canvas.

2

1. Lay aloft the port watch: she carries a total complement of 269, of which 200 are navy cadets.
2. Holystoning the decks: traditional methods of work go on side by side with the ability to understand and use some of the most advanced equipment. Gorch Fock II *is one of the smartest and best-equipped vessels of any description afloat today.*
3. Show a leg: the 200 cadets sleep in hammocks on two flats on the 'tween deck.
4. Captain Hans Engel, Master of Gorch Fock II *in the 1960s, eats with his crew.*

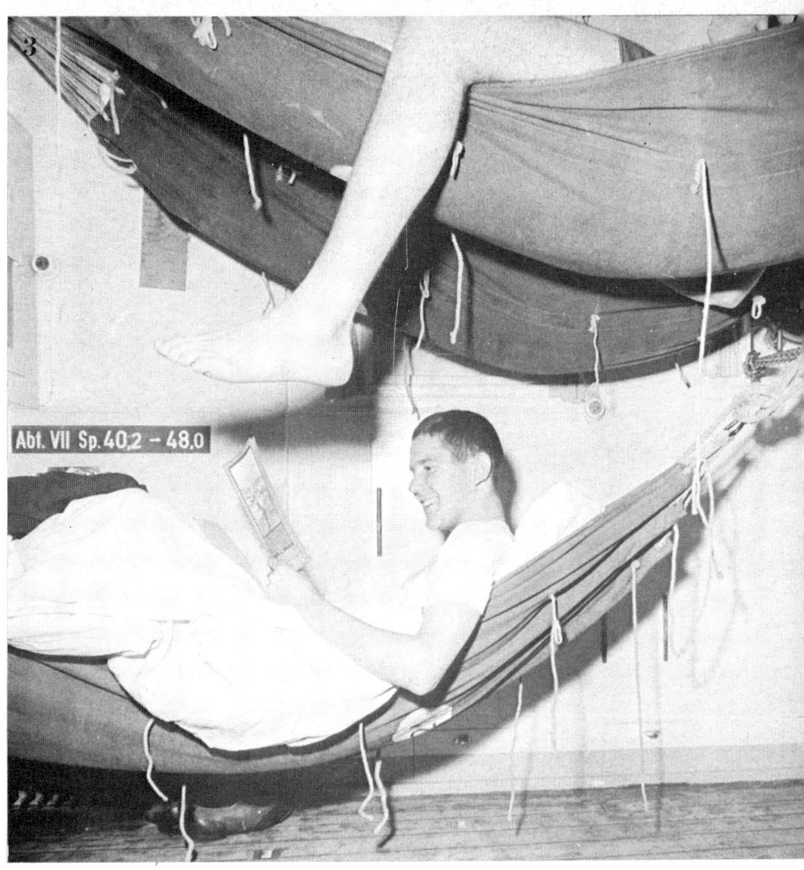

1. Eyes on the blades: boat drill in the approaches to Kiel.
2. Gorch Fock II *rig is similar to her Blohm and Voss-built sisters: double top sails, single to'gallants and royals. But like* Sagres II *she carries a divided spanker.*

ESMERALDA

The four masted barquentine *Esmeralda* of 3,445 tons displacement, is the training ship for officers and crews of the Chilean Navy.

The name *Esmeralda* was connected for the first time with the Chilean Navy in 1818 during the War of Independence when Commander O'Brien, captain of the 800 ton frigate *Lautaro*, armed with 44 guns, boarded the Spanish frigate *Esmeralda* which was blockading the port of Valparaiso. O'Brien died in the exploit, setting an heroic precedent for future generations of Chilean seamen.

The expedition to liberate Peru invaded El Callao a short time later and finally captured the *Esmeralda*, and she was commissioned into the Chilean Navy. In 1855, with the wars of independence over in South America, the Government of Chile ordered the construction of the second national ship which would bear the name *Esmeralda*. In 1889 a light cruiser was commissioned into the Chilean squadron, being the third national ship to bear the name *Esmeralda*. In 1898, the construction of the fourth *Esmeralda* was ordered. It was one of the most powerful cruisers of its time. In 1946, an anti-submarine frigate was named *Esmeralda* but transferred its name to the sailing vessel, when the latter was commissioned into the Chilean Navy in 1954.

This large barquentine is almost identical to the Spanish four masted, top sail barquentine *Juan Sebastian de Elcano* (see page 60). They were both designed by the British yacht designers and builders, Camper & Nicholson of Southampton. Both the Spanish and Chilean Naval Training Ships were built by the yard of Echevarrieta y Larriñaga of Cadiz, the *Juan Sebastian de Elcano* being launched in March, 1927. The *Esmeralda* was laid down in 1942 as the *Juan de Austria* for the Spanish Navy. However, when partially completed, she caught fire, with serious consequences. In 1951, under contract for the Chilean Government, she was rebuilt, launched in 1952 and commissioned in 1954.

The Chileans have always been keen on sail training, regarding it as an ideal medium for instructing their officers and men. The *Esmeralda*'s predecessors included *General Baquedano*, a steam auxiliary barque of 2,441 tons displacement and *Lautaro*, the ex Flying P Line steel, four masted barque *Priwall* built for F. Laeisz by Blohm & Voss in 1918. The German barque worked the west coast nitrate trade for her owners until 1926, when she was converted to classic sail training, having further accommodation added to enable her to train while trading. She was in Chile at the outbreak of the Second World War and Germany made a present of her to the Government of Chile, as a token of their long friendship. The Chilean Government handed her to the Navy which renamed her *Lautaro*.

The barquentine *Esmeralda* was built to carry on the tradition. Her total complement of over 300, includes 100 midshipmen and 90 seamen second class. She covers more miles than most training ships. When she visited Sydney in 1961, she was the largest sailing vessel to come to that port for 50 years.

Esmeralda took part in 'Operation Sail' in 1964, sailing into the port of New York in company with the rest of the fleet after the Tall Ships Race had ended at Bermuda.

The Chilean vessel carries no fore and aft rig on the foremast and is, therefore, a barquentine. Her fore and aft sails are secured to the mast by hoops. The bridge is on top of the deck house forward of the mainmast.

Esmeralda can manage 12 knots under power from her 1,500 hp Burmeister and Wain diesel engine. The exhaust from this is carried neatly up the lower jigger mast in the manner of the *Juan Sebastian de Elcano*. She carries four, 5·7 centimetre guns and her white hull is decorated by a large figurehead of Chile's national bird, the Condor.

OWNERS: Chilean Navy
DESIGNER: Camper & Nicholson
BUILDERS: Echevarrieta y Larrinaga
PLACE OF BUILDING: Cadiz
YEAR OF BUILDING: 1952
BUILT FOR: Chilean Navy
RIG: 4-masted barquentine
CONSTRUCTION: Steel
L.O.A. (feet): 308·8
DRAUGHT (feet): 28·5
BEAM (feet): 43·0
TONNAGE: 3,222 Disp.
PRESENT NATIONALITY: Chilean

Under power: the engine exhausts neatly through the cap of the jigger lower mast.

GLORIA

The three masted, 1,300 tons (T.M.) steel barque *Gloria*, is the most recent of the world's cadet barques. She is developed from the Blohm & Voss lines plan of 1933 which resulted in *Gorch Fock I*, now the Russian Navy's *Tovarishch*. As described earlier, these plans were used for *Eagle* (ex *Horst Wessel*) and the new *Gorch Fock II*, built in West Germany in 1958. *Gloria* is slightly smaller than these barques, as comparison with *Eagle* shows. The former's length overall is 249·3 ft, beam 34·8 ft, draft 15·9 ft, while *Eagle*'s same measurements are 294·3 ft, 39·04 ft and 17·1 ft.

Like *Eagle* and *Gorch Fock II*, *Gloria* carries divided topsails and single topgallants. Her spanker, like *Eagle*'s, is undivided, in contrast with *Gorch Fock II*.

Gloria's main distinguishing feature is her pronounced wheel-house and bridge at the forward end of the poop, between the main and mizzen masts. Another deck house extends from the forecastle to a point approximately half way between the main and mizzen masts. The structure does not extend to the ship's sides, leaving, therefore, side decks for access.

Gloria's complement is made up of a permanent crew of 50, including 9 officers and 60 cadets.

Although this Colombian training ship spends most of her time under sail, *Gloria*'s 530 hp diesel will give her 10·5 knots under power. Her training programme covers considerable distances. In 1969 she cruised from her home port of Cartagena to New York–Lisbon–Tripoli–Naples–Marseilles–Barcelona and back home. The next year, in a voyage of over 27,455 sea miles, she included both Australia and South Africa. Under Captain P. Uribe, she took part in 'Operation Sail' in 1972 at Kiel, following the Sail Training Association's races, Helsinki–Falsterbo, Sweden and the Solent, U.K. to Skagen, Denmark of that year. *Gloria* is well able to undertake extensive sea passages as she is designed to be able to remain at sea for 60 days without replenishing.

The cadet ships of South American countries certainly make good use of their vessels and the navies of Chile, Argentina and Colombia lead the world in this respect.

OWNERS: Colombian Navy
PLACE OF BUILDING: Bilbao
RIG: Barque
DRAUGHT (feet): 14·27
BEAM (feet): 34·75
DESIGNER: Colombian Navy
YEAR OF BUILDING: 1968
CONSTRUCTION: Steel
TONNAGE: 1,097 Disp.
BUILDERS: Astilleros Talleres Celaya S.A.
BUILT FOR: Colombian Navy
L.O.A. (feet): 249·3
PRESENT NATIONALITY: Colombian

Ghosting in light weather conditions.

DAR POMORZA

(ex *Princess Eitel Friedrich, Colbert, Pomorze*)

Dar Pomorza started life as the *Princess Eitel Friedrich*, the second training ship of the German School Ship Association.

The Association's first training ship was the full-rigged *Grossherzogin Elizabeth*, built in 1901 and one of the first training ships not designed for the dual role of training and carrying cargo. She is now the French Navy's accommodation vessel, *Duchesse Anne*, lying at Brest. The German School Ship Association's third vessel was the barque *Grossherzog Friedrich August* (built in 1914, now *Statsraad Lehmkuhl*, see page 64).

Princess Eitel Friedrich was launched from the Hamburg yard of Blohm & Voss in 1909. She is a little larger than her predecessor, *Grossherzogin Elizabeth*, being 1,561 as against 1,260 gross tons, but both ships have a similar profile, with a long poop stretching forward nearly to the mainmast.

The German School Ship Association trained both cadets and apprentices. The former were destined to be officers, the latter for the forecastle. None had much prospect of continuing their careers under sail, as there was a great demand for both officers and men in the rapidly expanding steam merchant service, but training under sail was considered essential, particularly for the officers.

The ship was handed over to France in 1918 at the end of the First World War. A similar fate befell her sister ship, now the *Statsraad Lehmkuhl* which was handed over to the British. Of the three, only the *Grossherzogin Elizabeth* stayed in German hands, to go to France in the end, as part of the reparation payment after the Second World War.

The French, however, had little use for *Princess Eitel Friedrich* and she was laid up in St. Nazaire until 1921. The Société Anonyme de Navigation Les Navires Ecoles Français, which used the four masted barque *Richelieu* as a training ship, then took her over and renamed her *Colbert*. However, her new owners failed to make use of her, even when they lost *Richelieu* by fire, in Baltimore Harbour, on 4th January, 1927.

Colbert was sold to the Baron de Forrest as a cruising yacht but was never used.

In 1929, the full-rigger was bought by the Polish State Sea School at Gdynia, after a public appeal. This ended her 11 years of idleness, caused by war and indecision. The Poles gave her her third and fourth change of name, renaming her *Pomorze*. She only rejoiced under the name of this Polish province during her tow to Poland, during which she was abandoned and nearly lost. When she arrived safely, she was given another change of name, becoming *Dar Pomorza* (*The Gift of the Province of Pomorze*). She replaced the aging barque *Lwow*, once the British-owned and built, iron, full-rigger *Chinsura*, which had been launched at Birkenhead in 1869.

From the end of January, 1930 until the outbreak of the Second World War, *Dar Pomorza* carried her 30 man crew and 150 cadets to many countries, including several Atlantic crossings, as part of her duties as the training ship of the Polish Merchant Service.

She sought safety in Swedish waters at the outbreak of the Second World War and remained in Stockholm until the war ended.

In 1946 she returned to her old duties and has carried them out successfully since that date, winning the Sail Training Association's Tall Ships Race, Isle of Wight to the Danish Skaw, in 1972. She also took part in 'Operation Sail' Kiel, during the sailing Olympics of that year, and competed successfully in the 1974 Copenhagen to Gdynia Race, organised by the Sail Training Association. Her master, Captain K. Jurkiewicz, has been with her for a considerable number of years and must rival the late Captain Hansen of *Danmark* as the longest serving master of a square-rigger.

OWNERS: Polish State Sea School
DESIGNER: Blohm & Voss
BUILDERS: Blohm & Voss
PLACE OF BUILDING: Hamburg
YEAR OF BUILDING: 1909
BUILT FOR: German Schoolship Association
RIG: Ship
CONSTRUCTION: Steel
L.O.A. (feet): 266·6
DRAUGHT (feet): 16·54
BEAM (feet): 41·0
TONNAGE: 1,784 T.M.
PRESENT NATIONALITY: Polish

1. In a fresh breeze during the Tall Ships Race in 1972.
2. Dar Pomorza *was winner of the 1972 Tall Ships Race.*

MIRCEA II

The steel barque *Mircea II* of 1,760 tons displacement, followed the first *Mircea* built at Blackwall in 1882.

The first *Mircea* was designed to accommodate a total of 80 permanent crew and cadets, while the second, built by Blohm & Voss of Hamburg, carries a complement of nearly 90 crew and 120 cadets.

Mircea II is a near sister ship to the line of Blohm & Voss barques that have proved to be such fine sail training ships. They are *Gorch Fock I* (1933, now the Russian *Tovarishch II*, see page 96); *Horst Wessel* (1936, now the U.S. Coast Guard training ship *Eagle*, see page 106); *Albert Leo Schlageter* (1938, now the Portuguese Naval training ship *Sagres II*, see page 118). Otmar Schauffelen in *Great Sailing Ships* records that another vessel of the same type, unnamed and unfinished, was used as a convenient way of disposing of unwanted gas shells in the Baltic, at the end of the Second World War.

Mircea II was completed by Blohm & Voss as ship number 519 in 1939, and sailed immediately for her base at the Merchant Marine Nautical College at Constanta, Romania.

After the war, she was taken over for a short time by the Russians, but was soon returned to her original owners and duties at Constanta.

She underwent a complete refit, including the installation of a new 1,100 hp diesel in 1966.

Mircea owes her name to the 14th-century Romanian prince, who by force of arms captured back from Turkey a vital length of coastline and so established Romania as a sea power. The figurehead of the barque is a representation of this Romanian hero.

OWNERS: Merchant Marine Nautical College
DESIGNER: Blohm & Voss
BUILDERS: Blohm & Voss
PLACE OF BUILDING: Hamburg
YEAR OF BUILDING: 1939
BUILT FOR: Merchant Marine Nautical College, Constanta
RIG: 3-masted barque
CONSTRUCTION: Steel
L.O.A. (feet): 239·5
DRAUGHT (feet): 16·5
BEAM (feet): 39·3
TONNAGE: 1,760 Disp.
PRESENT NATIONALITY: Rumanian

1. Overhauling the buntlines.
2. Under full sail.
3. Prince Mircea, the ship's figurehead. He gave Romania back her access to the sea after defeating the Turks in the 14th century.

Ships of the future

'In its essence life at sea has been always a healthy life, and part of that was owing to the very nature of the physical exertions required. I affirm with profound conviction that sailing-ship life is an excellent physical developer. I have repeatedly seen a delicate youngster brought on board by an anxious relative change out of all knowledge into a stout youth during a twelve months' voyage. I have never seen an apparently delicate boy break down under the conditions of the sea life of my time. They all improved. Moreover, any physical work intelligently done develops a special mentality; in this case it would be the sailor mentality; surely a valuable acquisition for a sea officer either in sail or steam.'

Joseph Conrad, 'Memorandum – On the Scheme for Fitting Out a Sailing Ship for the Purpose of Perfecting the Training of Merchant Service Officers Belonging to the Port of Liverpool', *Tales of Hearsay & Last Essays*, J. M. Dent, London, 1928, p. 71.

Many of the world's sail training ships are over 30 years old and a number are a good deal older than that. *Dar Pomorza*, for example, is 67, *Statsraad Lehmkuhl* 62 and *Kruzenstern* 50. This is not to say that they have exhausted the life before them, for good maintenance and use could prolong this almost indefinitely. *Dar Pomorza*, for example, is one of the most successful cadet ships afloat today. However, a new generation of training ships will have to be built, if the natural advantages of training under sail at sea are not to be lost.

Perhaps the best way of probing the future is to look again at the three types of ships – the classical sail training ship, built for training while trading; the cadet ship designed to provide sea training under sail for the Navy or the merchant service, and lastly the adventure training vessel, which applies part of the experience of the cadet ship to education and social welfare.

The Development of the Classical Sail Training Concept – Training While Trading

The demise of this form of training has been explored in the previous section and its former glory highlighted. For many, looking back, it was the only form of training that counted. Alex Hurst, one of the greatest authorities on the subject, explained that he thought there was no future now. 'It depends what you mean by training! Drill? Yes. Experience of a sort? Yes. But sail training is dead, and must remain dead, since it cannot become manifest in overmanned ships.'

Certainly it is difficult, even in the face of a real crisis in energy resources, to see the resurgence of the clipper in its old guise, driven shorthanded. Men have learnt easier ways of carrying cargo long distance and even the most adventurous would not come forward in sufficient numbers, paying sizeable premiums, to make a combination of classical and adventure sail training possible.

Today's ship would have to be modified to take advantage of the advances in material and sail technology. Thus the hard lessons would be altered, and the original 'stamp' lost. The routes that exploited the high latitudes and added so much to training may be no longer commercially essential – consider the effect of the building of the Panama Canal on the Chilean nitrate trade – and the press of time on a young man's sea career makes long voyages difficult to fit into any educational programme, particularly with the technical knowledge that has to be absorbed in the present atmosphere of educational and commercial impatience.

Despite brave ideas such as the possible recommissioning of *Peking* (ex *Arethusa*) for use in her original role, and the presence in reasonable shape of some of the survivors of the age, such as *Passat*, it is unlikely that classical sail training will be revived in its old guise, and at the turn of the 20th century, the last survivors of this experience will be no more, leaving only their records and reminiscences.

Nevertheless, there may yet be a future for carrying cargo under sail. Technical advances may modify the value of the old training, but may not kill it, as man understands better the value of natural experience and uses it more intelligently in education and personal development. A less crude and perhaps less physically demanding method of assimilating it can be achieved – this, wedded to a useful pollution-free task of cargo carrying, without the depletion of our non-renewable resources, is quite a goal.

Peking.

One of the pioneers of the new interest in carrying cargo under sail is Professor Wilhelm Prölss of the Institute of Naval Architecture at the University of Hamburg.

In his paper on the 'Economic Possibilities of Wind Propelled Merchant Ships', Professor Prölss recognised that the principal drawbacks of the last large sailing ships were their big crews, poor manoeuvrability, low average speeds and the lack of accessibility and small size of the cargo hatches. He proposed to overcome these difficulties by unstayed pivoted masts, carrying sail as an unbroken aerofoil, controlled remotely from the bridge. The sails, operated remotely, run on tracks in the fixed yards, the whole mast and sail array turning to achieve best results for the course steered. Reefing is inward toward the mast and, in a development of the rig, the sails would furl inside it.

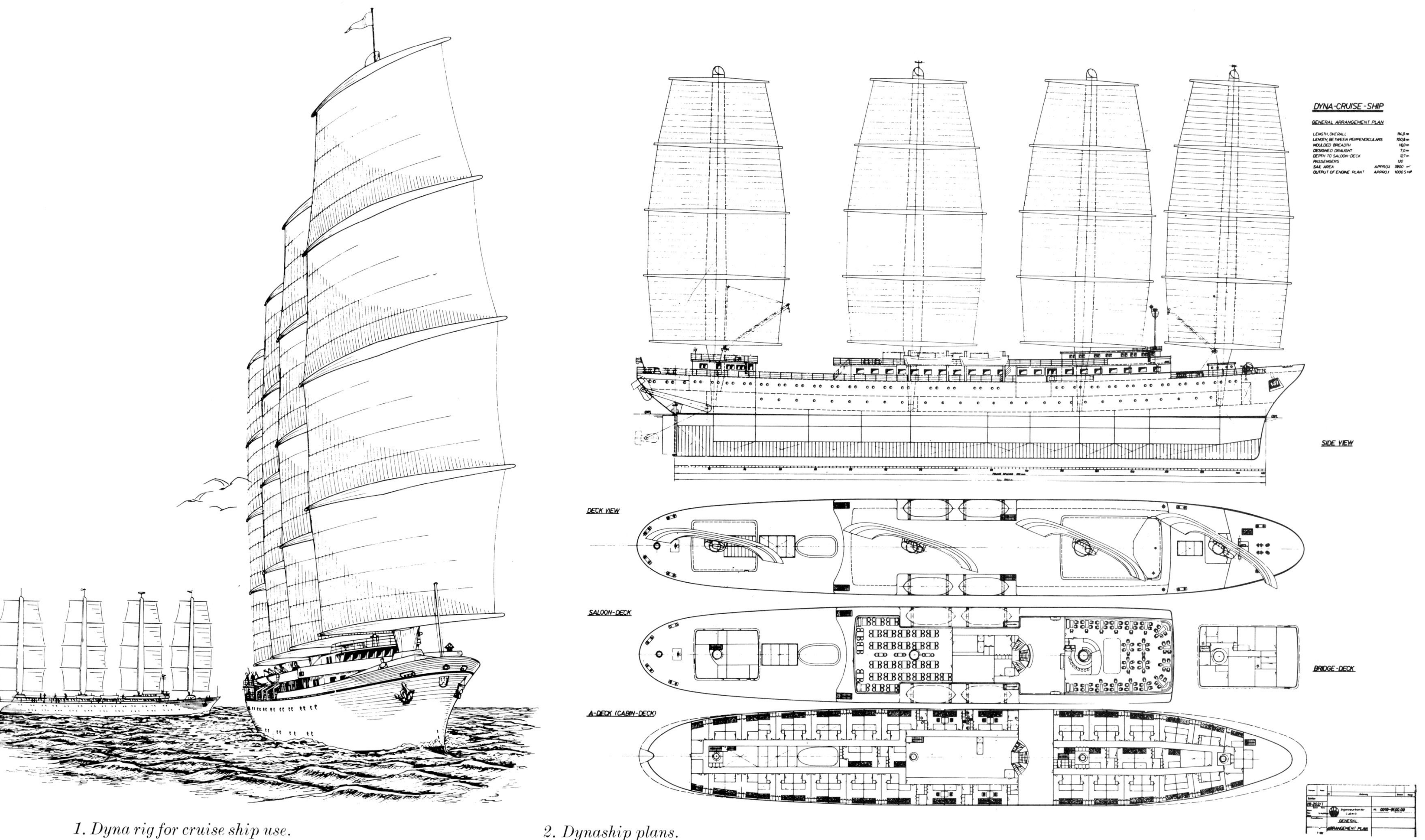

1. Dyna rig for cruise ship use.

2. Dynaship plans.

The British naval architect and yacht designer, Colin Mudie, M.R.I.N.A., who designed the brig *Royalist*, supplies another answer in the search for efficient methods of harnessing the winds to commercial purpose. He describes his ideas and the theory behind his revolutionary design:

'The traditional square rigger was something of a masterpiece of aerodynamic efficiency, in fact one authority has produced efficiency figures showing it comparing well with a twelve metre yacht if you take into account the whole range of sailing direction and not just windward work. What modern technology has to offer the sailing ship lies more in new materials and handling power possibilities together with the general aerodynamic clean up which they make possible. The modern sailing yacht Bermuda rig is no real guide in itself for it is aimed at absorbing the interest and energies of crews which are by commercial standards of prodigious size.

'The first assumption must be that we are talking more about very much reduced use of the fossil fuels rather than their complete elimination, certainly we are talking about vessels which will carry power units of one form or another. [Similar to the Dynaship in concept.] Reliable passage timing and indeed directed passage routing to make optimum use of the winds implies auxiliary engines. Modern thoughts for large sailing vessels tends towards twin engines both for "motor sailing" and for harbour manoeuvring. These auxiliary engines also conveniently offer power for normal shipboard facilities and also, more important, for the operation of power controls for the sailing units. With engines to make up minimum speeds the sailing units need not be designed for light weather, immediately cutting a great deal of cost and handling.

'One of the biggest developments in the world since the days of the square rigger is in modern lightweight materials and construction methods, principally due to aircraft technology. This lets us, in a stroke, get rid of the greatest bugbear of the square rigger, the windage of the rigging. The cantilevered, unsupported spar is now a real practical proposition. A second development which cannot be ignored is the high cost of tailor-made units compared with those of some kind of production line. Costs must have some relationship with those of the power units they are replacing (unlike for instance the modern sailing yacht where the sailing engine is likely to cost several times that of the diesel unit for the same performance). Another factor which must be taken into account is that it is the small ships rather than the very big ones which are least cost effective in a high fuel cost world. The biggest market for sailing units, therefore, is likely to be at the smaller end of the shipping range.'

Colin Mudie takes these ideas as the base for his design and has drawn a 'rig' that covers one, at least, of these requirements. He continues:

'First and foremost, each sailing unit is standardised, individual, and self-supporting. A balanced sail is essential, as it was in the square rigger, to cut down operating loads and the sailing unit therefore pivots at about the balance point for windward operation. To windward and reaching each unit can be power rotated into its operating position and the sail flow adjusted by outhauls to the twin booms. For reaching and running the booms are spread to open up a double sail something in the Ljungstrom manner to give enough area for an effective performance with the reduced relative wind speed. The sails can be operated by a number of methods but at the moment we think that an extending topmast, rather like a gunter, might be appropriate coupled with automatic slab reefing in the Chinese manner and brails for final furling. For repair and maintenance, hydraulic jacks could, for instance, hinge each unit down to deck level and for anything more complicated a unit could be completely replaced in a very short time by means of a dockside crane.'

This brief insight into a new development in the use of sail spells out real possibilities for commercial use. However, when such a ship is used for training, the picture is different. The need for men of courage and experience welded together into a team, each man individually competent in the arts of the mariner, disappears almost as quickly and as surely as on a modern motor vessel. The controls are protected inside and the master sits shielded in the warm centre of his web like a master magician with his cathode ray tube and dimmable dials.

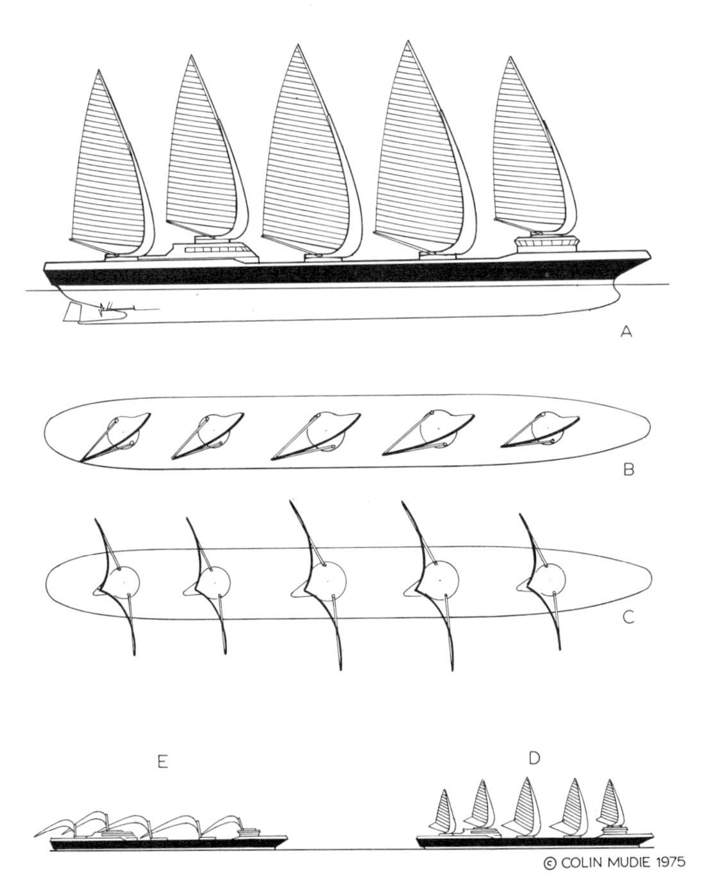

The Colin Mudie design: (a) A typical full sail profile. (b) The arrangement of slots for windward work. (c) The double sails extended for downwind sailing. (d) All sails reefed. (e) The units hinged down to the deck, for laying-up or general maintenance.

Except when things go wrong his crew need not go outside, or even put up with that typical mid-twentieth century smell, the oil and seawater odour of a modern merchant vessel.

This is, of course, praise indeed for both Professor Prölss' and Colin Mudie's ideas but it means that the opportunities of training in a natural environment are strictly limited. Even the sea is not close enough to help. This shortcoming could be rectified by adding the exploration of the environment to the purpose of the voyage, but this course is suited to the cadet or adventure training vessel, rather than to any future sailing merchant ship.

The Development of the Merchant Cadet Ship

Professor Prölss' dynaship rig and the Mudie rigs can also be used as power for a cadet vessel, and here simplicity of control is an advantage. The hours spent fighting the elements in sail changing and reefing, and engaged in the endless battle against chafe that dogs the life of the traditional vessel, can be spent learning the other arts of the mariner. Navigation and meteorology, for example, take on a new meaning, as the last ounce of power for the sails is borrowed secondhand from the sun. The extra time free from work also makes possible increased knowledge of the sea. Traditionally, the sailor has viewed the creatures that live in salt water with either passing interest or naked superstition – often both together. A large remora or sucking fish, for example, was said by sailors long past to attach itself to the keel to slow down a ship under full sail. Perhaps this is as good a reason for a tedious passage as any, but there is even today a traditional ignorance about the marine environment and this should be remedied in nautical colleges, a place being also found for such study while training at sea. This would add another dimension to a sailor's life which has been made more pedestrian by technical progress.

However, neither of the new concepts has yet to be given a full trial and the old and tested concept of the cadet ship is by no means dead. They still after all make up the largest class of sail training ship afloat today.

The Future of the Naval Cadet Ship
A short time ago the Hellenic Navy was interested in building a new cadet ship. Captain Mike Willoughby produced the magnificent, modern barque illustrated on page 145. She has a loaded displacement of 3,288 tons and has a sail area of 27,971 sq ft (2,599 sq m), which compares with the 21,011 sq ft (1,952 sq m) of the West German Navy's *Gorch Fock II*, the last training barque to be built. The new vessel was designed to have divided topgallants, unlike the German which carries divided topsails only.

1. Gorch Fock II.

The idea behind the barque, of course, was naval training, and she is very similar to *Gorch Fock II* (see page 126). She has a steel hull with five watertight bulkheads and there are two decks below the main, with raised forecastle and poop. The lower yards and topmasts are of steel, while the topgallant and royal yards are of timber.

The machinery is complex, as it was intended to carry engineer officer cadets. The barque is twin screw which is unusual in the larger ships of today and the main engines are air start, eight cylinder, 750 hp direct injection diesels that would give the training ship a speed of 11 knots under power. By comparison, her calculated performance under full sail in Force 7 is 16 knots.

The design includes three classrooms for radio communication, navigation and fleet operations, complete with electronic simulators. In addition there are engineering workshops and recreation spaces. Everything has been designed to give her a range of 10,000 miles.

With her complement of 80 permanent crew and 250 trainees, the new proposal represents perhaps the ultimate development of the traditional barque rig, allied to a hull that has many modern refinements. Consideration was even given to providing her with a bulbous bow.

The Greek Navy has, however, decided to postpone action on the ship for the time being, and it looks as if the Russians may be the next maritime nation to build a large, sailing cadet ship. They are reported to be interested in building two more barques, perhaps on the lines of the *Gorch Fock II*.

Such ships have many advantages over the smaller vessels in naval cadet training, for they can be equipped to provide the services of a technical university at sea. However, the smaller sailing vessels like the Swedish *Falken* and *Gladan* and the French *La Belle Poule* and *L'Etoile* help navies to provide a short sea-going phase under sail in their training curriculum.

Captain Willoughby's designs for a 68 ft (21 m) brigantine embody many of the proven ideas of the past, using modern materials and services. This beautiful little ship has a beam of 18 ft 10 in (5·8 m) and a loaded draught of 11 ft 3 in (3·4 m); her waterline length is 66 ft 8 in (20·3 m) and she displaces 155 tons. Accommodation is designed for two permanent crew and 20 trainees (see page 149).

2. La Belle Poule.

The British Joint Services have developed their own training vessels which are used for leadership training and for recreation. They are based on the standard Camper & Nicholson 55 ft (17 m) fibreglass yacht hull and fitted up for both work and relaxation. *Adventure*, the British Joint Services round-the-world race competitor,

Adventure.

is an example. Nine of these vessels have been built. Smaller Halcyon 27s and Contessa 31s complete the fleet.

The interior of the 55s has been designed to provide accommodation for 10 to 12 and the layout is to a standard laid down by the Ministry of Defence. There is a permanent skipper, though the vessels are generally maintained at the Joint Services Sailing Centre at Gosport by shore staff. When at sea, leadership and instruction is provided by a pool of experienced officers and N.C.O.s drawn from all three services.

The Development of the Merchant Navy Cadet Ship

Captain W. Wakeford, M.B.E., F.I.N., a past principal of the Southampton School of Navigation, is one of a number of British merchant navy educationalists who feel that sea training under sail in nautical colleges should be expanded. Even at Southampton, perhaps Britain's most advanced nautical college, a student would only have four hours afloat in any one academic week of 30 hours, that is 13·3 per cent. On this basis, a small training ship is used for about 370 hours of the 8,760 in a calendar year – in other words 4 per cent of the time. He felt that this was hardly satisfactory.

However, the situation has now changed and the governors have granted Southampton's 80 ft (24·4 m) ketch *Halcyon* a new lease of life. Captain Chris Phelan, the present principal, has well developed views on the value of sail training for the merchant navy. He believes that officers must have classroom knowledge but at the same time be practical men of high calibre and that some skills are better learnt while on the high seas. He sees power vessels having a place in this instruction, though emphasizing that training under sail is vital if 'we are not to lose the character of the people we need at sea'. He values close contact with the elements, commenting that it is easy to drive a super tanker without actually knowing what is going on 80 ft below. *Halcyon* will, therefore, be in operation for 31 weeks in 1976. Every entrant on the induction course will spend 5 days of that 13-week period aboard, training under sail. He sees this as an essential part of the development of the young seaman through living together, developing both responsibility, ingenuity and safety at sea, as well as practical seamanship and navigation.

Halcyon will also serve the local education authority, through their outdoor activities centre at Calshot, just across Southampton Water. This programme provides a multiple use and fulfils a broader educational need, which should expand during the summer months, during the Navigation School's holiday period.

If the United Kingdom is to reap the advantages of training under sail for the merchant navy in an economic way, there are two points of importance. The first is to increase the amount of sea time in the initial phase of a cadet's training and the second is to share the services of a purpose built sail training vessel between nautical schools. (This happens successfully today in Greece with the *Eugene Eugenides*.) There are 18 navigation schools in Britain, with a yearly intake of approximately 1,000 cadets, sufficient for an economic and educationally sound scheme using at least one ship. It may be possible to experiment in Britain with the help of the existing adventure training ships such as the *Sir Winston Churchill* and *Malcolm Miller* (see page 90). After all, their particular design was initially developed from merchant navy training proposals. Even assuming no dramatic alteration in the number of navigation schools in Britain, a new national board could co-ordinate the syllabi so that each training establishment had fair use of the new cadet training vessel.

Such a scheme could be undertaken not only in Britain but also in other maritime countries of the world. In Japan, for example, nautical colleges are served by the two four masted barques *Nippon Maru* and *Kaiwo Maru* (see page 122).

A national approach to nautical training gives an opportunity to provide sail and power vessels on a shared basis, and could either give new life to existing ships through better use, or provide the incentive for new building. The need for the sailing cadet ship, both for navy and merchant navy training, must increase in the future if the needs of the mariner are properly studied. Reducing skills to that of shore-based repetitive industry will have the same consequences as on land, disenchantment followed by creeping industrial paralysis.

The arguments for the need to add breadth to the professional naval officer's knowledge of his chosen environment are just as true for those who look forward to a commercial life at sea. This wider dimension should be provided while undergoing training at sea under sail.

The Adventure Training Ships

Perhaps because of experience gained from the other forms of training under sail, the brightest future of all lies ahead

PROPOSED GREEK TRAINING SHIP

PROPOSED OWNERS: Hellenic Navy
PLACE OF BUILDING: none
RIG: Barque
DRAUGHT (feet): 17
BEAM (feet): 41

DESIGNER: Capt. Mike Willoughby
YEAR OF BUILDING: none
CONSTRUCTION: Steel
TONNAGE: 3288 Disp.

BUILDERS: subject to international tender
BUILT FOR: Hellenic Navy
L.O.A. (feet): 288·7
PROPOSED NATIONALITY: Greek

of the adventure training ship. The idea was slow to develop, and the use of the sea for adventure training, with its associated self-examination and development, is still being explored.

The composition of the course is of paramount importance and comes before the design or modification of the vessel, as John Hamilton, Race Director designate of the Sail Training Association and a past mate of the *Captain Scott*, explained:

'It has been proved that it is more exciting and has more impact if, at the start of a cruise, the crew join by tender from the shore rather than just climb aboard with the vessel in a marina or alongside a quay. I believe this is because the launch acts as a bridge between the normal, civilised life which they are leaving and the new life they are to experience for a week, fortnight, or whatever. This difference is made even more apparent as they step aboard; masts tower above them, mysterious ropes are coiled up everywhere, there is the smell peculiar to sailing ships, the hum of ancillary equipment, pipes, fans, pumps, lifebelts, and so on – all a different world.'

The permanent crew is of great importance and this has to be born in mind particularly when designing a new scheme. Again John Hamilton explains:

'A big factor is the special position of the adults on board. In this new, strange world, the adult crew of the vessel are the ones who know what is going on and how to cope with the many strange equipments and procedures. In the widest terms, they are the ones who can re-establish the bridge back to normalcy and get the youngster back to the "safe" world he knows. Therefore, unless the adult is a complete fool, he has the respect of the young crew automatically and can build on that respect to weld his watch or the ship's company into a team.

'With a group of clever adults, the weakness in a youngster can be countered by finding out something at which they excel over the others and this can be highlighted so that the remainder of the watch can see, therefore, where there may have been disparagement, it is replaced by respect.

'The team spirit of the watch is nearly always fostered, perhaps a quiet remark about its performance compared with another watch, the hoisting of a sail that much faster, a few more miles covered in the four-hour watch – this teamwork nearly always leads to a heightened comradeship, perhaps the kind of comradeship the youngster has never met in his home life. Brought about by a group meeting and sharing a potentially dangerous or exciting situation and mastering it together.'

After selecting the course and considering the sort of people required to make it work, there is the need for the right ship. There are basically two schools of thought. The first which could be called traditional, regards the fore and aft rig as regrettable and its developments as necessary only because such design enables the ship to go to sea with an inexperienced crew and only a short pre-training period. This means a day and a half in the case of the Sail Training Association's schooners, for example, while the German Navy cadet barque *Gorch Fock II* takes three weeks alongside for such preparation. Supporters of the traditional idea argue that real training comes from the use at sea of the squaresails, with work aloft rather than on deck. A proper discipline aboard, therefore, is vital to the safe handling and necessary maintenance of the ship. Mike Willoughby's brigantine is an example of how a ship can be designed to suit this school of thought and yet go to sea relatively quickly, without a long period of harbour drill.

The Canadians, through Sea Adventure Ltd., of Calgary, Alberta, are exploring a similar concept with a proposal for a 130 ft (39·48 m) barquentine from the board of James D. Rosborough of Halifax, Nova Scotia.

A unique idea that follows this first school, but for other reasons, is that put forward by the British company Cinemarine Ltd. The directors of this concern have plans to build a 120 ft (36·48 m) sailing hull that can be rigged and modified as a film ship able to cover 300 years of British maritime history. She is designed to be used as a sail training and ecological study vessel when not required for her primary film-making duties. She is to be built at Napier's Wharf on the Thames, near the heart of London, where Brunel built his *Great Eastern*. In addition, it is rumoured that a French concern is considering building a small wooden barque for sail training purposes, though no details are to hand.

The traditional school is, therefore, still strong and can point with force at a lively catalogue of varied and present interest, built on lessons of the past.

The other school emphasises the advances that have been made in sailing technology and highlights the value of training gained from the proper handling of a really efficient sailing machine. Speed wedded to close windward ability with an easily driven hull are matters of moment. No really large vessels of over 100 ft (30·48 m) have been built with this in mind so far, but the British Ocean Youth Club, the Italian Navy, and others have developed this form of adventure training for their own particular purposes using the modern form of sailing yacht of 70–80 ft (21–24 m).

The traditional ideas have an advantage in that the rig suits a load-carrying hull and good accommodation can be provided in a comparatively short waterline length. Such a ship is also likely to have a longer life as a training ship, for her performance through the water is of much less consequence than that of vessels designed to the more modern specifications of the second school of thought. The 'old fashioned' look does not date, for it continues to work effectively. Modifications are made when more efficient or economic alternatives to rig or superficial equipment are available. The basic vessel alters little.

On the other hand, the fast sailing vessel concept presents a challenge with each major advance in sail and hull technology. These improvements happen continuously because of the quest for speed and efficient harnessing of the wind for yachts. On these finer vessels, the gear is lighter, the vessels smaller and so the consequence of a mistake is potentially less disastrous, and the tighter discipline and organisation on deck of the larger ship is less essential. It is seldom necessary to go aloft.

Whatever school of thought is followed, the success of adventure training is difficult to measure exactly. Trainees,

1. Captain Scott.

2. Expedition on St. Kilda from Captain Scott. Commander Clark (bearded) with the expedition's officer, John Hinde.

their parents, firms, education authorities, and others are often pleased and surprised by the results. As some theoretical psychologists and educationalists find the 'sea change' difficult to quantify, they also find the value of such training awkward to accept. This has made anything but development by experience and experiment, carried out by those running the various adventure schemes, nearly impossible.

However, the course must dictate the design of vessel, and if science is unable to play a constructive part in the course building side, it has an undoubted effect on naval architecture.

The design of the adventure training vessel of the traditionalists will therefore advance slowly. That of the second school of thought will progress at almost the same speed course wise, but will need to allow for later change and improvement to be built into the vessel in order to keep pace with the progress of sailing technology.

The real advance in adventure training and design of vessels devoted to such purpose comes from adding further purpose to self-discovery and the advantages of working together.

The *Captain Scott*'s ecological cruise (30) in January, 1975, even though done with adults outside the normal age range and off the north-west coast of Scotland in very bad weather and at a poor time of year for study, showed how this might be achieved. The schooner, which is designed to operate a month's cruise and includes carefully worked-up shore expeditions as routine, was able to introduce her crew to a smattering of oceanography at sea, carry out a necessary tree planting expedition on a Hebridean island, study the habits of the Manx shearwater and undertake a deer count.

But the ship had not been designed for these extra activities, and difficulties arose. In company with other adventure training vessels, the rig was designed to provide work, which was time and energy consuming, and this severely limited the period available for the short oceanographic programme. Ashore, time spent preparing for normal 'outward bound' land expeditions had to be curtailed in order to undertake a programme of studies. Normally the work-up of gradually extended trekking, climbing and living in the open culminates in the final expedition. *Captain Scott* fortunately possessed for her land expeditions a large store of special clothing and equipment with suitable drying facilities to cope with Scotland's winter wilds and summer wet, but she had no laboratory or other facilities for oceanography or for carrying, for example, seedling trees. Ecological study was limited to making use of such time the working of the ship would allow and such equipment and stores that could be easily brought aboard.

It is difficult to adapt a ship built for a particular purpose to another, without major structural change which is not usually economically feasible. It is, thus, of paramount importance to work out both the purpose of the course and of the vessel with great care. The four years that were taken to raise the money by British public subscription for the schooner *Sir Winston Churchill* was

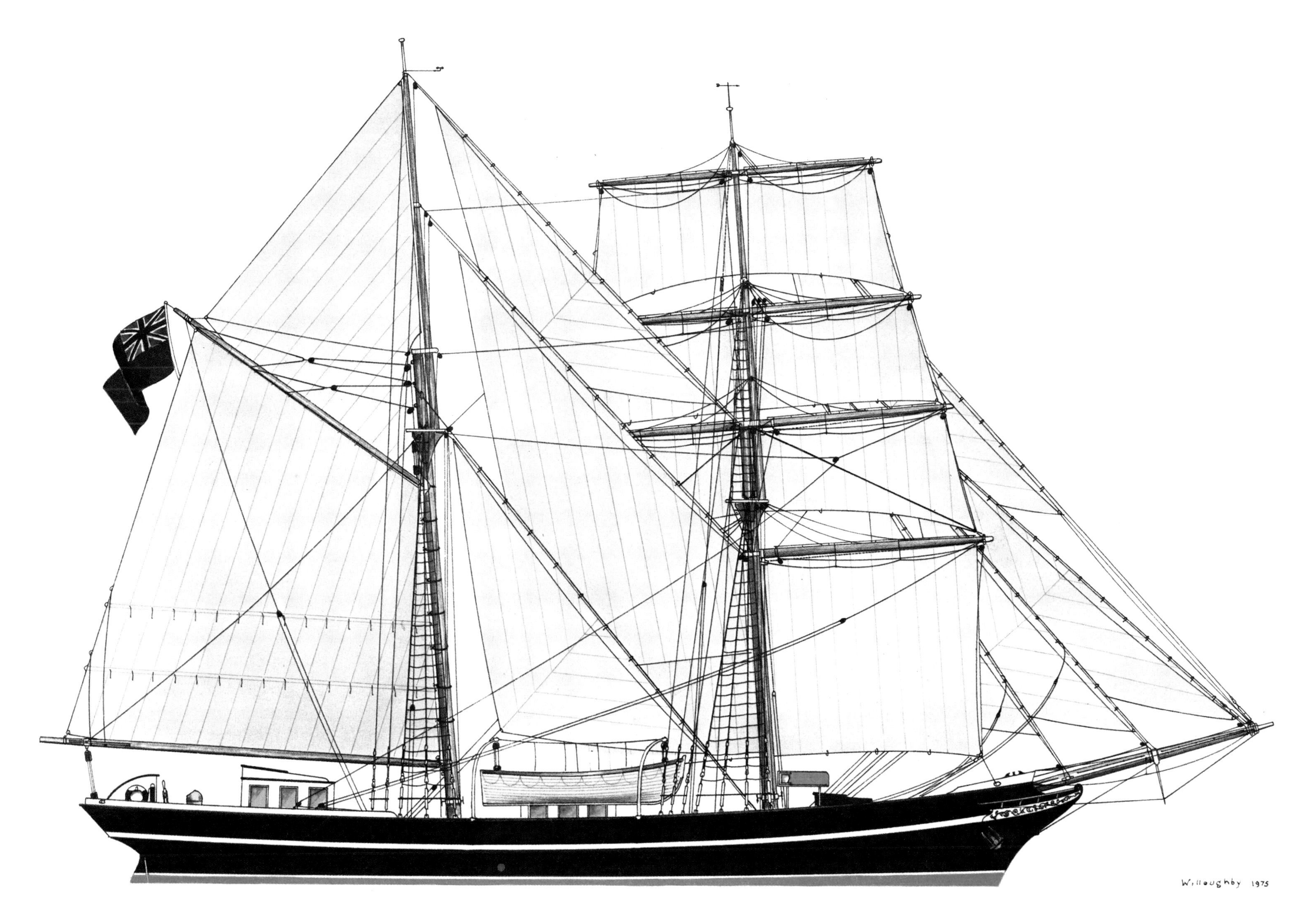

PROPOSED BRIGANTINE

PROPOSED OWNER: Mr. James Simpson
PLACE OF BUILDING: none
RIG: Brigantine
DRAUGHT (feet): 13·8
BEAM (feet): 18·8

DESIGNER: Capt. Mike Willoughby
YEAR OF BUILDING: none
CONSTRUCTION: Wood
TONNAGE: 155 Disp.

BUILDERS: none
BUILT FOR: Mr. James Simpson
L.O.A. (feet): 68
PROPOSED NATIONALITY: British

just time enough to design the two-week cruise and the schooner to carry it out efficiently.

Adding to personal knowledge of the ocean environment may be reason enough to create a new course and new ships, particularly when grafted onto the stem of adventure training under sail. But the course can also be designed to be useful in the broader sense by increasing man's knowledge generally through survey and observation. Such development adds to the value of the idea, as its purpose can be easily seen by the public and those aboard, and can help the scheme itself economically. It is not difficult to give an example of how this could be achieved today.

The future of energy supplies for Britain, Scandinavia and western Europe depend in the short and medium term on the exploration and exploitation of the oil potential discovered in the North and Celtic seas and the western approaches to the Channel. These areas are valued at the same time as a valuable resource for fisheries. They also provide a vital habitat for other forms of wild life, particularly sea birds and seals. It would bring into question the long term value of man's attempt to harness this shorter term energy potential, if at the same time, he destroyed or seriously damaged this natural resource. This can be done easily by failing to understand the conflicts that exist between the ocean and coastal ecosystems and man's own demands. It is important therefore that the ecology of these areas be well understood. This could be achieved by sponsoring surveys and later 'monitoring cruises' under sail while exploitation of oil reserves is going on. The work could be done by an adventure training or indeed a cadet vessel under sail and provide valuable additional knowledge, training dimension and revenue to the course itself.

Such a ship could also provide further opportunities for limited numbers of navy and merchant service cadets who could help officer the ship and so experience their first active and responsible role. The scientific work, where trained personnel are essential, might form part of a degree or post graduate study at universities. In addition, as a third training opportunity, the work of the sailing of the ship and its daily tasks could well be undertaken by trainees engaged in adventure training.

Such a ship, with its peculiar responsibilities, would fit this new role best if purpose designed and built. For apart from additional accommodation to the normal for specialist instructors and scientists, further storage and oceanographic equipment spaces, laboratory facilities and specialised ship's boats would have to be provided aboard. The capacity of *Captain Scott* would have to be wedded to the capability of *Westward* and to this blend added a new rig to make further time for the new onus on scientific work and navigational instruction. Both the Dynaship and Colin Mudie's designs provide an interesting answer that particularly fit this last requirement, though, as has been emphasised, modifying the experience and value of working a more traditionally rigged ship under sail. The design of the course would have to take this into account by providing new work and a new challenge at sea.

The study of the marine environment is not, of course, the only possible addition to adventure training courses under sail. There are others that can fulfil an outside need and at the same time develop skill and personality by providing an opportunity of responding to challenge.

Medium sized adventure training ships carry large crews and so are able to undertake a number of jobs in a short time, particularly in remote areas that would require major effort by a few, or otherwise need an expensive expedition. The repair or excavation of ancient monuments and other suitable archaeological work, the building of observation posts for island and coastal ornithological study, are examples. Skilled instructors and leaders can be signed on for particular tasks, seconded perhaps by specialist universities and teaching colleges. On longer cruises, work of social significance could be undertaken in seaports, the activity depending on the part of the world, local need and request. The land based British Outward Bound Trust has pioneered this sort of adventure training work in cities.

Training under sail, with or without the additional development described above, has a role in reclaiming those who have become dependent on drugs or alcohol – the Danish training schooner *Fulton* has made progress in this field. The necessity of having to work at a variety of new and different tasks aboard ship provides a respite from normal everyday pressures and an opportunity for self-reassessment that can help when life is taken up again ashore.

The development of adventure training calls, therefore, for the development of the new and specialised vessels but, at the same time older vessels, including those of national significance historically, can play a part, as long as both the courses and the waters are suitable.

It is obvious that the sea has a great deal more to offer man if the challenge and the study of both ocean and coastal waters are properly harnessed. The sailing ship is the most rewarding method of achieving this and a new series of courses for Navy, merchant navy and general education should be designed to take advantage of such a natural resource, especially when demands on land are accelerating. Such progress will not only give national and international reward, but will stimulate new sailing ship design, as well as prolonging the life of those that are doing such excellent work today.

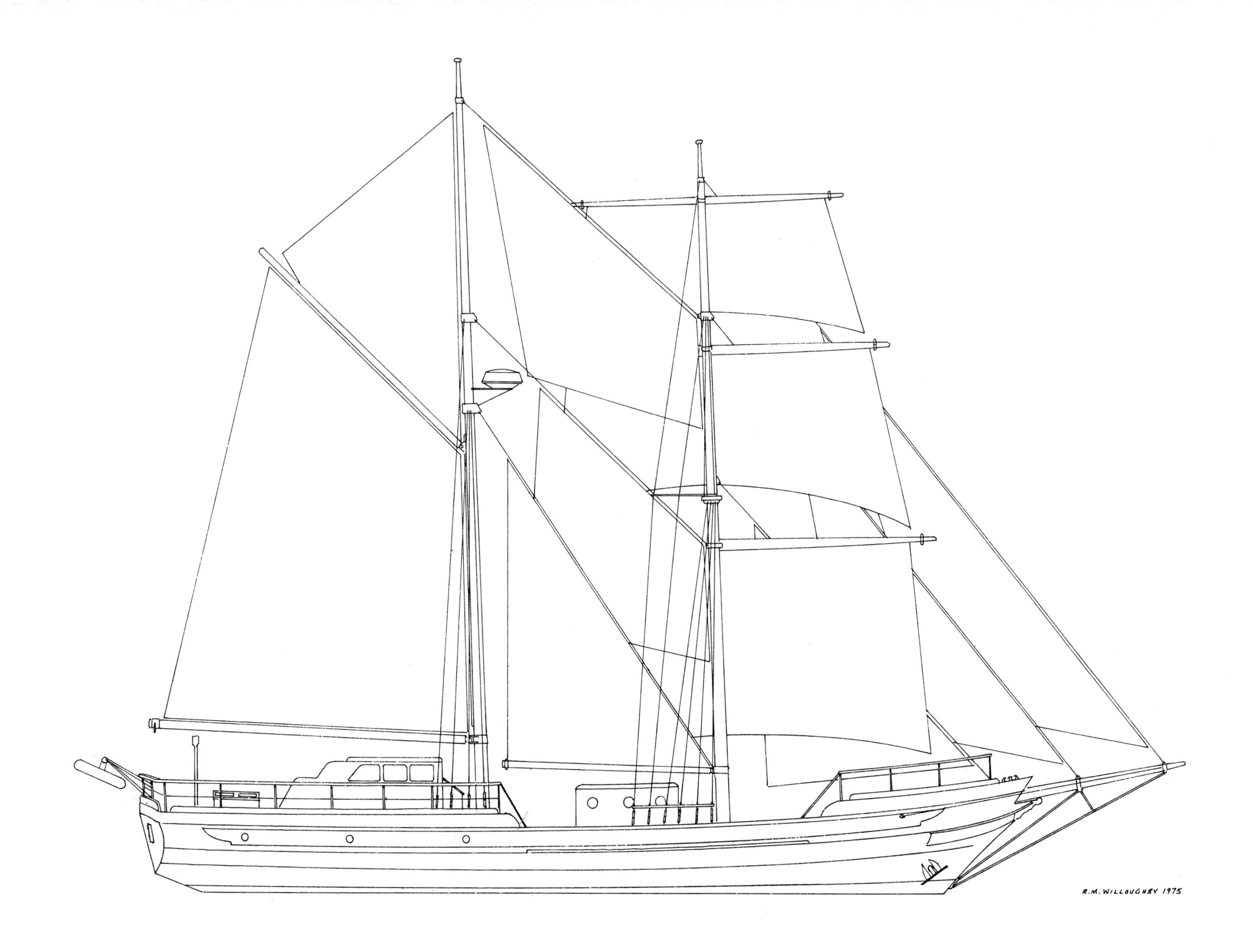

BRENDAN

PROPOSED OWNERS: Irish Sail Training Committee
DESIGNER: John Tyrrell
BUILDERS: Tyrrell's of Arklow
PLACE OF BUILDING: Arklow, Eire
YEAR OF BUILDING: 1976
BUILT FOR: Irish Sail Training Committee
RIG: Brigantine
CONSTRUCTION: Wood
L.O.A. (feet): 84
DRAUGHT (feet): 9·5
BEAM (feet): 18·8
TONNAGE: 120 Disp.
PROPOSED NATIONALITY: Irish

Appendix of ships' rigs

Stay Sails

1. *Flying Jib*
2. *Outer Jib*
3. *Inner Jib*
4. *Fore Topmast Staysail*
5. *Fore Staysail*
6. *Jib Topsail*
7. *Jib*

7a. *Genoa Jib*

14. *Royal Staysail*
15. *Topgallant Staysail*
16. *Topmast Staysail*
17. *Staysail*

} *Named after the masts abaft the sail. i.e. 16. Main Topmast Staysail.*

Boom Sails

18. *Foresail*
19. *Mainsail*
20. *Mizzen*
24. *Spanker*

Gaff Sails

21. *Fore Topsail*
22. *Main Topsail*
23. *Mizzen Topsail*

Yard Sails

8. *Skysail*
9. *Royal*
10. *Topgallant*
11. *Upper Topsail*
12. *Lower Topsail*
13. *Course*
25. *Cro'jack*

} *Named after the mast on which the sail is set. i.e. 10. Fore Topgallant.*

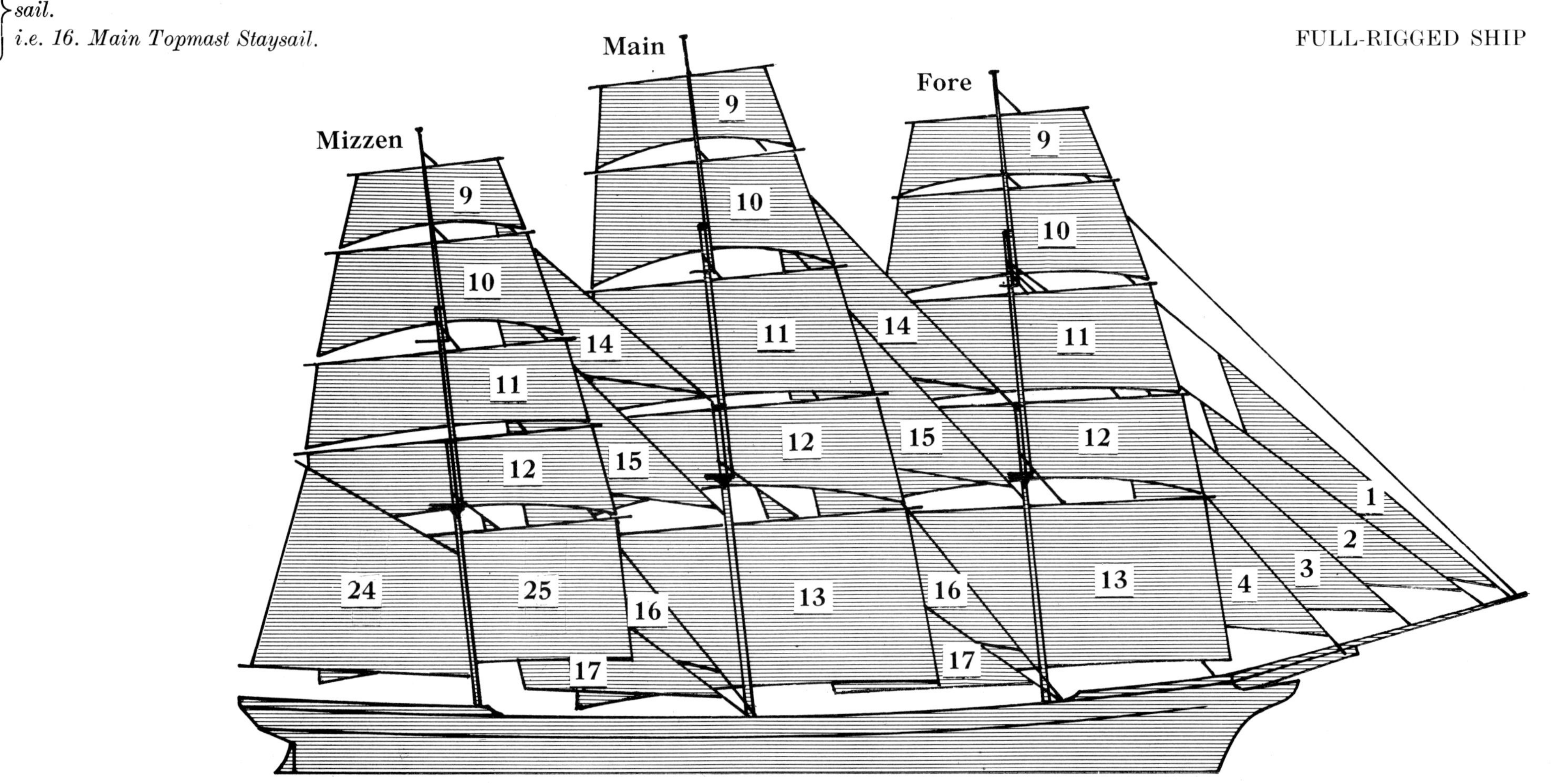

FULL-RIGGED SHIP

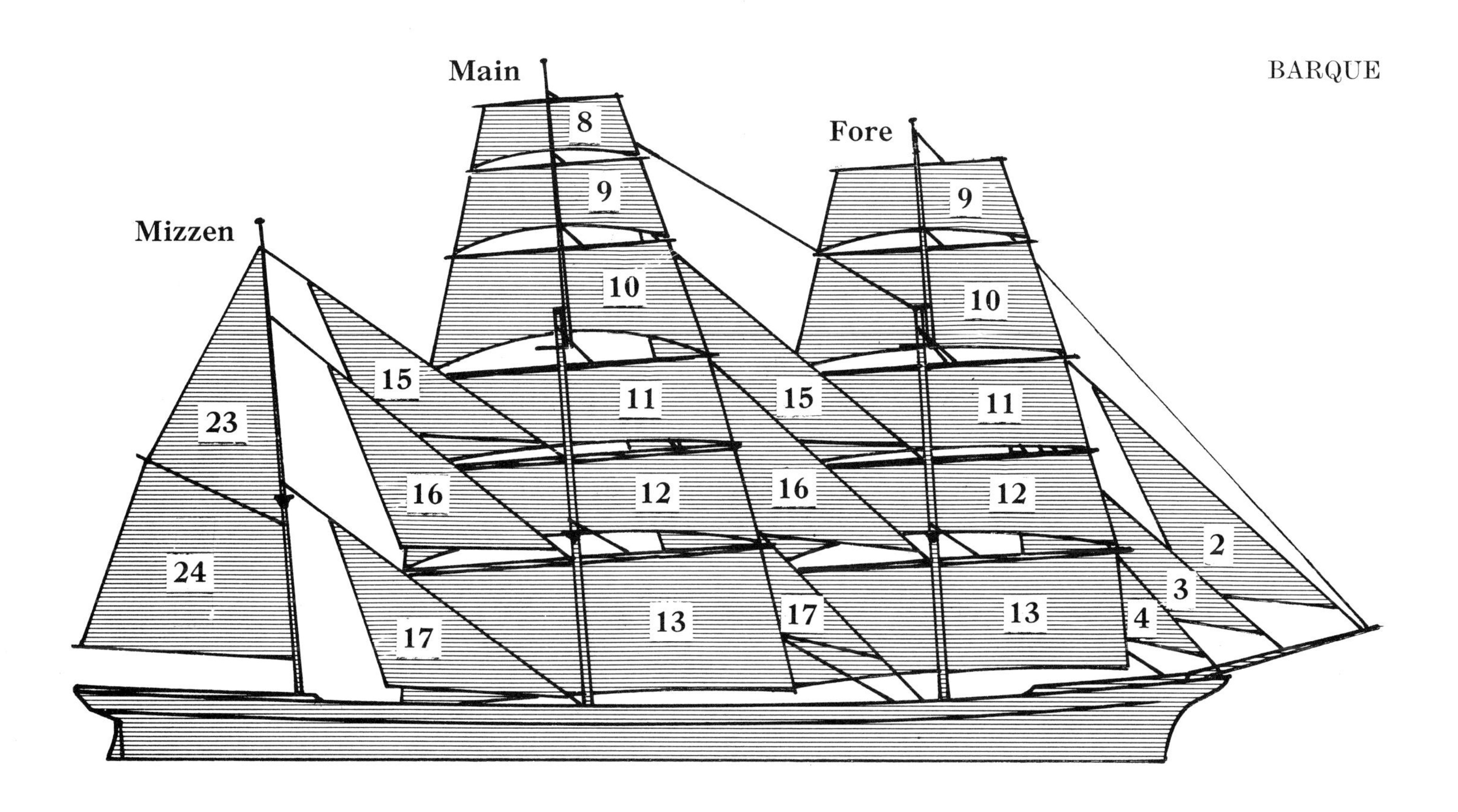
BARQUE
Main
Fore
Mizzen
8
9
9
10
10
15
11
15
11
23
16
12
16
12
2
24
3
13
17
13
4
17

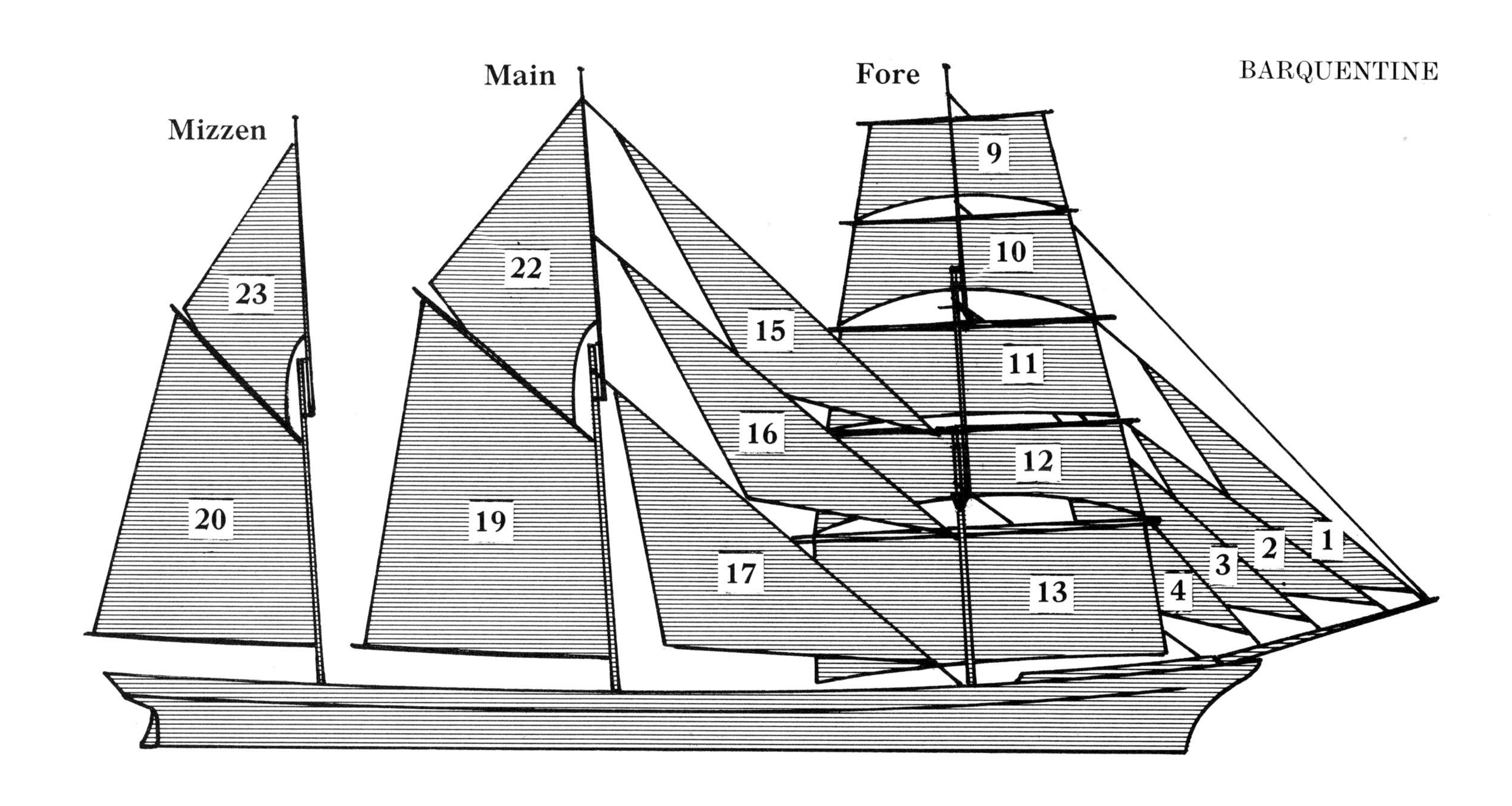
BARQUENTINE
Main
Fore
Mizzen
9
10
22
23
15
11
16
12
20
19
1
2
3
17
13
4

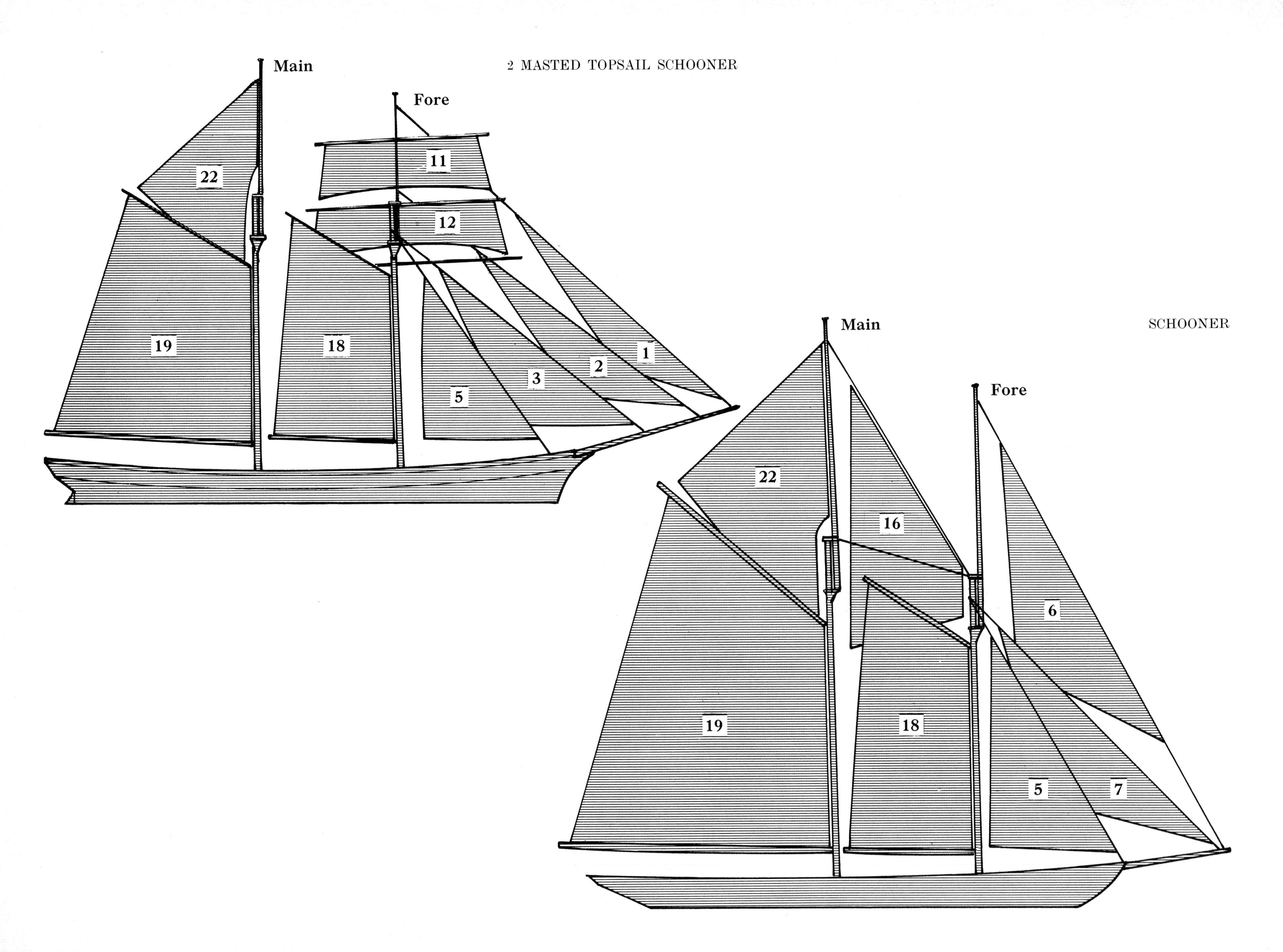

2 MASTED TOPSAIL SCHOONER
Main
Fore
22
11
12
19
18
5
3
2
1
SCHOONER
Main
Fore
22
16
6
19
18
5
7

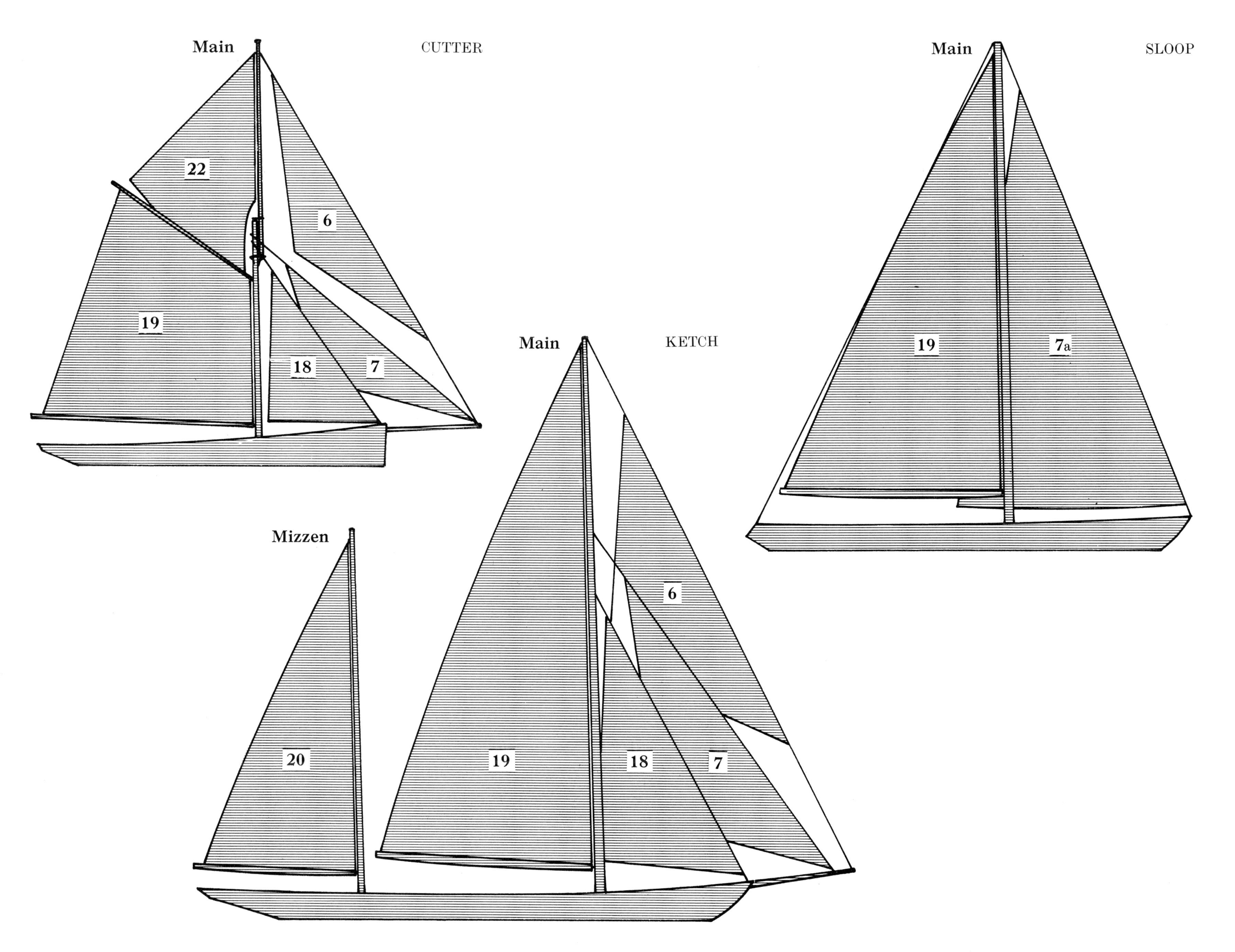
Main
CUTTER
22
6
19
18
7
Main
SLOOP
19
7a
Main
KETCH
Mizzen
6
20
19
18
7

Bibliography

Anderson, John — *The Last Survivors in Sail* — Percival Marshall & Co. Ltd., London, 1948.

Bock, Bruno — *Windjammer – Lübeck Kiel 1972* — Koehlers Verlagsgesellschaft Mbtt., Herford, W. Germany, 1972.

Clark, Commander Victor, D.S.C. — *On the Wind of a Dream* — Hutchinson, London, 1960.

Colton, J. Ferrell — *Last of the Square-Rigged Ships* — G. P. Putnam & Sons, New York, 1937.

Conrad, Joseph — *Tales of Hearsay & Last Essays* — J. M. Dent, London, 1928.

Crumlin-Pedersen, Ole — *Skonnerten Fulton Af Marstal* — Vikingeskibshallen, Roskilde, Denmark, 1970,

Derby, W. L. H. — *The Tall Ships Pass* — David & Charles Reprints, Newton Abbot, England, 1970 (1st Edition, Jonathan Cape, 1937).

Hansen, Hans Jürgen & Wundshammer, Benno — *Windjammer Parade* — Gerhard Stalling Verlag, Oldenburg, W. Germany, 1972.

Hansen, Hans Jürgen — *Heis Die Segel!* — Gerhard Stalling Verlag, Oldenburg, W. Germany, 1974.

Hurst, A. A. — *Square Riggers, The Final Epoch* — Teredo Books, Sussex, England, 1972.

Hurst, A. A. — *The Sailing School Ships* — Hemmel Locke, Ltd., London, 1962.

Jones, Clement — *Sea Trading and Sea Training* — Edward Arnold & Co. London 1936

Lubbock, Basil — *Colonial Clippers* — James Brown & Son, Glasgow, Scotland, 1924.

Lund, Kaj — *Training Vessels Under Sail* — Skandinavisk Bogforlag A/S, Odense, 1969.

Meyer, Jurgen — *Hamburgs Segelschiffe, 1795–1945* — Verlag Egon Heinemann, W. Germany, 1971.

Norton, William I. — *Eagle Seamanship – Square Rigger Sailing* *U.S. Coastguard Academy Manual* — M. Evans & Co. Inc., New York, 1969.

Norton, William I. — *Eagle Ventures* — M. Evans & Co. Inc., New York, 1969.

Pearse, Ronald — *The Last of a Glorious Era* — Syren & Shipping Ltd., London, 1934.

Prölss, Professor Wilhelm — *On the Economic Possibilities of Wind Propelled Merchant Ships* — Institut für Schiffban der Universität Hamburg, 1967.

Schäuffelen, Otmar — *Great Sailing Ships* — Adlard Coles, London, 1969.

Stackelberg, Corvettencapitän Freiherr Hans Von — *Rahsegler Im Rennen – Reisen und Ragatten der Gorch Fock* — Verlag Duburger, Vücherzentrale, Flensburg, W. Germany, 1965.

Underhill, Harold A. — *Sail Training and Cadet Ships* — Brown, Son & Ferguson, Glasgow, 1956.

Underhill, Harold A. — *Deep Water Sail* — Brown, Son & Ferguson, Glasgow, 1952.

Underhill, Harold A. — *Masting & Rigging* — Brown, Son & Ferguson, Glasgow, 1946.

Underhill, Harold A. — *Sailing Ship Rigs & Rigging* — Brown, Son & Ferguson, Glasgow, 1938.

Villiers, Alan J., D.S.C. — *The Making of a Sailor* — George Routledge & Sons Ltd., London, 1938.

Villiers, Alan J., D.S.C. — *The Cruise of the Conrad* — Hodder & Stoughton, London, 1937.

Villiers, Alan J., D.S.C. — *The Last of the Wind Ships* — George Routledge & Sons Ltd., London, 1935.

Villiers, Alan J., D.S.C. — *Voyaging with the wind* — National Maritime Museum. HMSO London 1975

Magazines, journals and papers

All Hands — Magazine of the Old Cadets Association & the School of Navigation, University of Southampton, England.

Cruising Opportunities — An R.Y.A. Publication, G11/73, London.

Lloyds 100 A1 Bulletin — No. 15, 1965, Lloyds Register, London.

The Log of Mystic Seaport — Quarterly Journal of the Marine Historical Association, Incorporated Mystic Seaport, Conn., U.S.A.

Mariners Mirror — Quarterly Journal of the Society for Nautical Research, London.

Proceedings — Monthly Journal of the United States Naval Institute, Annapolis, U.S.A.

Sail — Yearly Journal of the Sail Training Association, Bosham, Sussex, England.

Sea Breezes — Monthly Magazine of Ships and the Sea, Liverpool, England.

Index

176029
359
D844T
Drummond, M.
Tall ships: the
world of sail training
WITHDRAWN